SCHOOL DESIGN MATTERS

Presenting qualitative and quantitative findings from the unique, multi-disciplinary project, *Design Matters?*, this timely book explores the complex relationship between school design and practice to consider how environmental aspects impact on the day-to-day perceptions, actions and behaviours of pupils, teachers, leaders and professionals within the school community.

Exploring debates and issues from a number of different professional and academic perspectives, *School Design Matters* results from a rich collaboration between schools, architects, engineers, educationalists and policymakers to consider what an inspiring teaching and learning environment might look like. Case studies and first-hand student and teacher experience allow analysis of the ways in which environmental factors might transform pedagogy, shape patterns of leadership, improve student engagement and enhance social interactions within and beyond the school community. Experts in their fields, authors acknowledge the significance of sociocultural contexts, reference relevant policy, and tackle the tensions, dilemmas and contradictions which frequently arise as schools and professionals in the design and construction sectors collaborate in the creation of buildings which fulfil the needs of diverse, invested parties.

Offering a uniquely holistic approach to understanding the ways in which design may contribute, shape and mediate teaching and learning, this comprehensive text will be essential reading for educationalists, architects, policymakers and professionals involved in the design, construction and use of school buildings.

Harry Daniels is Professor of Education in the Department of Education, University of Oxford, UK.

Andrew Stables is Professor Emeritus at the University of Roehampton and a Senior Researcher with the International Semiotics Institute.

Hau Ming Tse is a Research Fellow in the Department of Education, University of Oxford, UK.

Sarah Cox is a Researcher in the Department of Education at the University of Oxford, UK.

SCHOOL DESIGN MATTERS

How School Design Relates to the Practice and Experience of Schooling

Harry Daniels, Andrew Stables, Hau Ming Tse and Sarah Cox

Routledge
Taylor & Francis Group

LONDON AND NEW YORK

First published 2019
by Routledge
2 Park Square, Milton Park, Abingdon, Oxon OX14 4RN

and by Routledge
52 Vanderbilt Avenue, New York, NY 10017

Routledge is an imprint of the Taylor & Francis Group, an informa business

© 2019 Harry Daniels, Andrew Stables, Hau Ming Tse and Sarah Cox

The right of Harry Daniels, Andrew Stables, Hau Ming Tse and Sarah Cox to be identified as authors of this work has been asserted by them in accordance with sections 77 and 78 of the Copyright, Designs and Patents Act 1988.

All rights reserved. No part of this book may be reprinted or reproduced or utilised in any form or by any electronic, mechanical, or other means, now known or hereafter invented, including photocopying and recording, or in any information storage or retrieval system, without permission in writing from the publishers.

Trademark notice: Product or corporate names may be trademarks or registered trademarks, and are used only for identification and explanation without intent to infringe.

British Library Cataloguing-in-Publication Data
A catalogue record for this book is available from the British Library

Library of Congress Cataloging-in-Publication Data
A catalog record for this book has been requested

ISBN: 978-1-138-28010-6 (hbk)
ISBN: 978-1-138-28011-3 (pbk)
ISBN: 978-1-315-27241-2 (ebk)

Typeset in Bembo
by Apex CoVantage, LLC
Printed and bound by CPI Group (UK) Ltd, Croydon, CR0 4YY

CONTENTS

List of figures — *vi*
List of tables — *viii*
Acknowledgements — *ix*
About the authors — *x*
Foreword: design obviously matters — *xi*
Peter Clegg
Foreword — *xiii*
Pamela Woolner

1	Introduction: school design matters?	1
2	School design: what do we know?	7
3	The aims, scope and method of the *Design Matters?* project	33
4	Design as a social practice	59
5	Design and practice	89
6	The experience of new-build schools	119
7	Changing schools	139
8	What matters about design?	167

Appendix: School connectedness questionnaire — *186*
Index — *187*

FIGURES

1.1	Overall schema of *Design Matters?*	2
2.1	BSF case study school	8
2.2	Wider consultation	9
2.3	BSF case study school	10
2.4	Control options	11
2.5	Traditional classroom	15
2.6	Teaching and learning environment with a mixed economy of space	15
2.7	Example of open-plan space with ICT	19
2.8	Floorplan of radical BSF design	20
2.9	Special needs school with central atrium space	24
3.1	School display work	37
3.2	School display work	38
3.3	School display work	38
3.4	School display work	39
3.5	School display work	39
3.6	School display work	40
3.7	Overlapping semiotic codes of schooling	43
3.8	*Design Matters?* methodology	47
3.9	Case study school	48
3.10	Case study school	49
3.11	Case study school	49
3.12	Case study school	50
3.13	Case study school	50
4.1	The general structure of coordination	63
4.2	The general structure of cooperation	63
4.3	The general structure of communication	63
4.4	Historical forms of work	64
4.5	Educational model for new learning spaces	71
4.6	School C4 concept exterior	75

Figures **vii**

4.7	School C4 learning zone sectional perspective	78
4.8	School C4 concept plan	79
5.1	Bexley Academy	91
5.2	Perspective view of initial design concept	92
5.3	Perspective view of typical village cluster	93
5.4	Locality C, Schools C1–C4 and Comparator School CC1	94
5.5	School C1 design concept	98
5.6	School C1 design concept	99
5.7	School C1 design in practice	100
5.8	School C1 design in practice	100
5.9	School C2 design concept	102
5.10	School C2 design concept	103
5.11	School C2 design in practice – time point 1	104
5.12	School C3 design concept	106
5.13	School C3 design concept	107
5.14	School C3 design in practice	108
5.15	School C3 design in practice	108
5.16	School C4 design concept	110
5.17	School C4 design concept	110
5.18	School C4 design in practice	111
5.19	School C4 design in practice	111
5.20	School C4 design in practice	112
5.21	The production and deployment of the building as an artefact	114
5.22	Diagram of sustainability	115
5.23	Waddington's depiction of an epigenetic landscape	116
5.24	The flexible underpinning structure in Waddington's epigenetic landscape	117
6.1	*Design Matters?* methodology	123
6.2	Time point 1: Proportion of primary students that referred to each category, by type of school when asked 'What do you think will be different about your new school?'	124
6.3	Time point 2: Proportion of secondary students that referred to each category, by type of school when asked 'List in order of priority, the most important spaces to you?'	125
6.4	Time point 1-3: How secondary students' responses to preferences for categories changed through time in School B1	125
6.5	Time point 1-3: How secondary students' responses to preferences for categories changed through time in School D1	126
6.6	Time point 1-4: How secondary students' responses to preferences for categories changed through time in School A1 as the design and educational practices changed	127
7.1	Overall, connectedness by measurement occasion	150
7.2	School connectedness scores by measurement occasion and secondary school	151
7.3	School connectedness scores by measurement occasion and type of secondary school	152
7.4	School connectedness scores by measurement occasion and school modality group	153
7.5	Connectedness scores in relation to design/practice alignment/conflict	154
7.6	Jane's mean school connectedness score over time	158
7.7	John's mean school connectedness score over time	160
8.1	Concept diagram of case study school	178
8.2	Concept drawing of special school	181

TABLES

4.1 Vision through to occupancy: stage model. Example: Locality C 67
4.2 Summary of main points for each stage of the vision/design/build process 82
7.1 Examples of elements of the coding frame for classification of spaces as designed and framing over use of space as envisaged in the design 146

ACKNOWLEDGEMENTS

We would like to thank the Arts and Humanities Research Council (AH/J019924/1) for their funding and support. We have had the pleasure of collaborating with a large number of schools, educators, design professionals, engineers and academics who have all been exceptionally generous with their time. Most importantly, students from all over the country have been highly engaged with our research and have willingly shared their views and concerns that have been central to this research.

This project has lasted five years and over that time we have had the privilege of working with talented researchers, including Susannah Learoyd-Smith, Lorena Ortega Ferrand, Rebecca Tracz, Sarah Roper, Victoria Read, Grace Murkett, Irem Alici, Emma Powell, Eszter Saghy and Zening Yang. We would particularly like to thank Adolfo Tanzi Neto for his special contribution to our research project. We would also like to thank our steering committee – Gert Biesta, Phil Blinston, Alison Clark, Peter Clegg, Ian Grosvenor, Rob Hannan and John Jenkins – for their insightful guidance.

Cover image credits:
Top left: © BAM Construction Ltd
Top right: © HKS
Middle right: © Martine Hamilton Knight
Bottom right: © HKS

ABOUT THE AUTHORS

Professor Harry Daniels is Professor of Education in the Department of Education, University of Oxford and holds Professorships at The Australian Catholic University, Griffith University, Moscow State University and Kansai University. He specialises in the development of post-Vygotskian and activity theories and studies learning across a wide range of domains from these perspectives. His current research focuses on the relationship between design and practice and exclusion from school.

Professor Andrew Stables is Emeritus Professor of Education and Philosophy at the University of Roehampton, London, having previously worked at the universities of Bath and Swansea and with visiting positions at Ghent, Chaiyi (Taiwan) and Oxford. He is particularly interested in the implications of semiotic philosophy for understanding educational and other social practices.

Hau Ming Tse is a Research Fellow in the Department of Education at the University of Oxford, invited expert for the Department of Education, UK and member of the Technical Advisory Group for the OECD Centre for Effective Learning Environments. She was an Associate Director at David Chipperfield Architects until 2007. Selected projects include the Hepworth Gallery, Wakefield and the Headquarters of BBC Scotland, Glasgow. Her current research focuses on productive points of interaction and innovation between theory and practice in learning environments. *Design Matters?* funded by the AHRC examined the complex relationship between design and pedagogic practice in some of the most challenging secondary schools in the UK

Sarah Cox is a researcher in the Department of Education, University of Oxford and has supported many educational research projects. She was an integral member of the AHRC funded *Design Matters?* team and contributed to all aspects of the project.

FOREWORD: DESIGN OBVIOUSLY MATTERS

Peter Clegg

Design *obviously* matters. My biased view comes as an architect with 25 years experience in the design of secondary schools. We simply would not be in this business if we did not think we could have a profound impact on the support of a school's social structure and pedagogical ambitions. But this enquiry has made us think profoundly, and has allowed us time for reflection on what has worked and what has caused problems.

Recently we have lived through the third major phase of school building in just over a century. The first one emerged from the 1902 Education Act, which made provision for secondary education to the age of 14. The second one responded to the 1944 Act and also the post-war baby boom. And the third responded to a further rise in the school age population, combined with a recognition that much of the original post-war buildings were in a dire state of repair.

Added to this was the political motivation from the Blair/Brown/Blunkett government at the turn of the century that education needed social and economic transformation in a way that had not been seen for more than half a century. So began, first, the Academies Programme and, in 2004, the £55bn 15-year Building Schools for the Future (BSF) programme to transform every school in the country. The extraordinary ambitions of this programme were cut short by a combination of the 2008 financial crash and the incoming Conservative government in 2010. The pendulum swung in a completely opposite direction with minimal expenditure on only those schools in desperate need, plus new Free schools promoted by those with individual curricular ambitions to supposedly generate greater choice for the parent-consumer.

Whatever side of the political football game of school design you support, it is extraordinary that more research has not been done into one of the great social experiments of the twenty-first century. This study goes a long way to filling that gap. BSF was introduced with a requirement that there should be continuous learning through post-occupancy evaluation, but very little of any substance took place, and if it did, it was confined to measurable data on issues such as energy performance or attainment and attendance. Daniels, Stables and Tse bring to the table a much broader range of evaluation techniques from semiotics and spatial psychology. They use in-depth studies to analyse the way pupils and teachers have responded to the new spaces they have been given. They show how different approaches to management and leadership result in very different uses and often costly adaptation of buildings. They talk of schools as 'activity systems' and develop a concept of 'school connectedness' when talking, in particular, about the dramatic change from primary to secondary environment that is so often accompanied by a reduction in performance, attainment and social well-being. And perhaps most importantly they highlight the successes when the ambitions of the headteacher and the designers are aligned, and the pitfalls of misalignment.

BSF was very consciously an invitation to the professions of educators and architects to re-examine in detail the very nature of a school. The goal was nothing less than the *transformation* of education. There is no doubt that the breadth and indulgence of the enquiry lead to inefficiency and may have encouraged architectural arrogance. But the very process of enquiry resulted in much higher standards generally. Improved acoustical performance, ventilation rates, better equipped kitchens and larger dining areas, increased space standards, a very different approach to WC provision, improved energy performance and of course more extensive use of IT. In many cases these standards have been reduced again as the pendulum swung and austerity kicked in, but many have are still with us and have changed the nature of a school compared to 20 years ago.

Fourteen years after BSF we are left trying to build schools to budgets which are less than 50 per cent of what they were 10 years ago, and to space standards which take us back to where we were 20 years ago. And the 2017 National Audit Office study, in words that could be a repeat of what was being said 20 years earlier talks of the 'parlous state of the school building estate' compounded by problems with rising demand for places (again at record levels) and problems with the delivery of capital projects.

Some of the more interesting lessons learned from this research come in the form of 'notings' in the final chapter. These provide a commentary on how things have changed over the last 20 years in ways we might not have noticed. Dramatic increase in security, rapid changes in technology infrastructure, observations on a necessary sense of belonging and engagement, all make the incidental learning from this study valuable to future building designers. We need to learn from this very comprehensive piece of research how we can face the next challenge in the provision of better school places.

FOREWORD

Pamela Woolner

The publication of the results of the *Design Matters?* project comes at a time of increasing international recognition of the educational importance of the physical environment. School design certainly matters, in the UK and beyond.

This interest in school space plays out differently across the world. In the global south, the overarching concern is the provision of sufficient adequate facilities. However, many industrialising nations are also trying to adapt school designs to accommodate their particular climatic and cultural needs. In the United States, research from a perspective of educational equity centres on assessing the negative impact of inadequate buildings, which are disproportionately found in poor neighbourhoods. For the last ten years, the Australian government has been funding a flurry of school-building, intended to enable 'innovation', but also to support the construction industry through the global economic downturn. Meanwhile, many European countries are building new schools to accommodate future global citizens, but while holding onto local variations and national idiosyncrasies.

School design in the UK, on which the project centred, has seen particularly dramatic developments since the beginning of the century. Building Schools for the Future (BSF), running from 2004 to 2011, was a government programme explicitly concerned to 'transform' education through the provision of individually designed premises, tailored to the particular needs and aspirations of the school community. Discontinued after a change of government, it has been replaced, amid much rhetoric and polemic, by the Priority School Building Programme (PSBP). This programme aims at rebuilding schools, not transforming education, with schools built rapidly to standardised designs, based on corridors and enclosed classrooms (although toilet design has stayed innovative).

Yet, even as subsequent British governments have derided BSF as wastefully expensive, they have completely accepted that some schools are not physically fit for purpose and require rebuilding. And, in schools, people are talking about teaching and learning space in a way that didn't happen when I was a student in the 1970s and 1980s, or in the 1990s when I was teaching. Even as I began researching school buildings in the early years of BSF, the excitement about the physical learning environment didn't extend to educational researchers. Many seemed bemused by my interest, despite the research, then expanding rapidly, into the virtual learning environment. In contrast, there is now an awareness of the built environment of education that extends across policymakers, researchers, school staff, students and parents.

However, the next step is to move from awareness of school space to build understandings of the complex relationships that develop between physical setting and educational aims, processes and outcomes. This will enable research to inform policy and practice into the future, countering the rhetoric that often dominates discussion of school buildings. We have (mostly) moved beyond the question that I was always being asked in the early years of the last decade – 'Does the building affect learning?' Now it is mainly recognised that appropriate questions concern how the physical space relates to the values, intentions and practices of the inhabitants; that education is more than learning; and that we require ideas from a range of perspectives to develop our educational spaces and use them well.

This book, reporting research conducted at a key moment in UK school design, is an important contribution to the discussion. Whether you come to this volume for an introduction to this exciting field or for the next instalment in the expanding interdisciplinary conversation, there is much to interest and inform.

Credit: © Martine Hamilton Knight

1
INTRODUCTION
School design matters?

The *Design Matters?* project investigated the relationships between design, practice and experience of secondary schools built in England in the early 2000s under the *Building Schools for the Future* (BSF) initiative and the Academies Programme (see facing page for an example). We studied a sample of these new-build schools within a decade of their inception, alongside comparator schools in the same locality. Our interest was in the ways in which design influenced the day-to-day experiences of school communities, with a particular emphasis on the transitions between different learning environments.

Design Matters? was a unique multi-disciplinary collaboration between schools, architects, engineers, educationalists and policymakers. The history and methodology of the project will be described more fully in Chapter 2. In brief, our aim was to extend beyond typical approaches to post-occupancy evaluation conducted on the performance of new buildings in order to gain insights into the effects of new designs on the lived experiences of members of school communities. We had a particular interest in the effects of design on pupils transferring from primary school as shown in Figure 1.1. We followed cohorts of students from Year 6 (ages 10–11), the final year of primary school in England, to the end of Year 8 (ages 12–13) through whatever changes in pedagogic approach and leadership took place. We developed a range of innovative, non-directive approaches to data capture, in order to arrive at the richest possible account through a range of first person ('I/We') and third person ('She/He/They') perspectives in these successive occupations.

The theoretical perspectives we brought to bear were inevitably influenced by the lead investigators' previous work, particularly Daniels' work on schools as cultural historical activity systems, and Stables' work on education and semiotics.

A central issue in the *Design Matters?* project was that of the relationship between the school context and perceptions and actions of the individuals who study and teach within it. The field abounds with descriptors such as 'sociocultural psychology', 'cultural historical activity theory', each of which has been defined with great care. However, confusions persist alongside what still appear to be genuine differences of emphasis. As Wertsch et al. (1995: 11) argue, they all attempt to explicate the 'relationships between human action, on the one hand, and the cultural, institutional, and historical situations in which this action occurs on the other'. Our challenge brought questions about the nature of context into focus. A key figure in the development of cultural historical activity theory is Michael Cole (1996) who distinguishes between two understandings of the word 'context'. The first is roughly equivalent to the term

Design Matters?

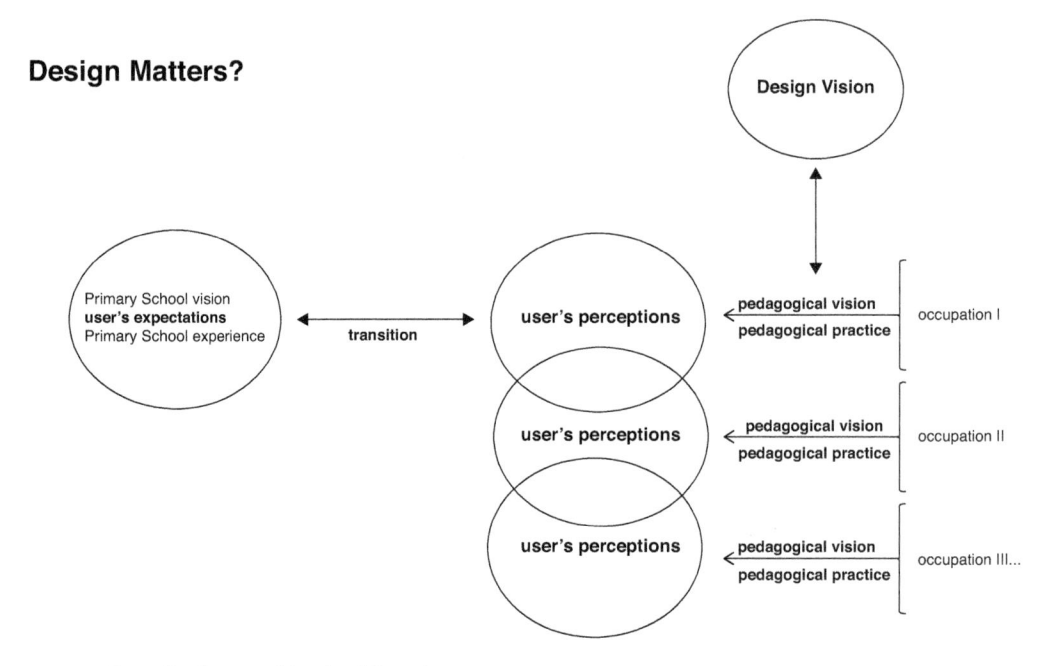

FIGURE 1.1 Overall schema of *Design Matters?*

'environment' and refers to a set of circumstances, separate from the individual, with which the individual interacts and which are said to influence the individual in various ways (Cole, 2003). Use of this understanding can lead to studies of how a context, in our case a school design context, influences action within a school. In the second, understanding individual and context are seen as mutually constitutive. In the words of the *Oxford English Dictionary*, context is 'the connected whole that gives coherence to its parts', a definition which has strong affinities to the Latin term 'contextere', or to weave together. When used in this way, the ability to segment child and the context is problematic, but an analytic distinction that depends upon a large, perhaps unaccountably large, set of factors operating in bi-directionally over time in an active process of framing that can be unravelled in an instant (Cole, 2003). It is this second understanding of context which underpins the various approaches to the study of human activity that have been derived from the work of the Russian social theorist, L.S. Vygotsky. This notion of context is important in our work because it renders the tactic of parsing independent variables and identifying single factor effects (such as school design) problematic.

In activity-oriented approaches, conceptual isolations between individual, objective world and activity are avoided. Vygotsky (1987) offers a dynamic and wide-ranging model that explains the process of internalisation of semiotically mediated social forces. Even apparently the most individual and autonomous actions are situated in a context which must itself be viewed as an active participant in the structuring of their activities. This understanding of the active making of context in which design provided the tools for the active construction of school context was a key understanding at the inception of the project.

Stables has also long been interested in semiotic mediation in schooling, focusing on perspectives from the history of semiotics. From this he has, through a series of publications, attempted to develop a distinctive 'edusemiotic' perspective (e.g. Stables, 2016; Stables et al., 2018).

From this semiotic perspective, our surroundings do not determine how we respond to them. Consider how the agoraphobe or the claustrophobe would react to a small dark room or an open field. Each of us is predisposed to attach certain significances to certain aspects of the places in which we find ourselves. This is not merely a matter of psychological or emotional trauma, our assumptions about social class and individual identity are also important. Each of us is strongly impelled by considerations of what 'people like us' and 'people like me' do and where we belong. For example, this is witnessed in one school context in the statement of a young student who when interviewed about his newly built school stated that 'this school is too posh for us'.

Living is semiotic engagement (Stables, 2006), a process of engaging with signs, each of which has a different significance for us. Signs operate in space and time, and when spaces are occupied and used at particular times they become places. We are always in places: places are the sites for the events that constitute our lives. The environment is a collection of significations not merely a mass of entities to which we are indifferent. Selection and values lie behind what we notice, find important, and like or dislike about where we are. This is not merely true of the human world: consider that the same blade of grass in a field may be a snack for a cow but a pathway for an ant. Semioticians refer to the environment of any organism as its umwelt. Each organism has its way of negotiating its environment: its innenwelt. Two other semiotic concepts that are relevant to a discussion of the role of school design are those of lebenswelt (the human cultural world: a term not exclusive to semiotics) and semiosphere, the totality of all significations (Lotman, 2005). On this account, a school is a particular umwelt located within a particular lebenswelt (by calling the place a school, we are giving it particular cultural meaning), to which all relevant actors – teacher, students, parents, ancillary staff, even visitors – react according to their particular innenwelt. In this sense, the innenwelt may be compared to Pierre Bourdieu's concept of the habitus: the set of dispositions that collectively determine how we respond to a particular situation, or, in Bourdieu's terms, locate ourselves within a particular 'field' (Bourdieu, 1993). The semiotic account only differs from Bourdieu's in being less explicitly sociological, though it does not deny the value of sociological categories, and in taking a strongly organic view of the individual and their response to their environment, that renders a structural account as always less than complete. For example, social class might be understood as a powerful influence, but even studying the intersection of class with other structural factors such as gender, age or ethnicity falls short of a complete understanding of an individual's life world.

Taken together, these sets of perspectives provided the analytical framework for *Design Matters?*

Overall, as the project progressed, we came to agree that one issue above all was predominating: that of the relationship of design to practice. This is both complex and at times difficult to understand, as we found some of our sample schools underwent several changes of occupation, and that different occupiers interacted with the design in different ways. What seems beyond doubt, however, is that design influences practice, but it does not simply determine or 'cause' it.

Design Matters? would not have been possible without the invaluable support of many schools, teachers and students and others in the architectural and educational professions, and of other academics with similar interests to ourselves. Above all, we wish to thank the AHRC for supporting the project with grant AH/J011924/1.

References

Bourdieu, P. (1993) *The Field of Cultural Production.* Cambridge: Polity Press.
Cole, M. (1996) *Culture in Mind.* Cambridge, MA: Harvard University Press.
Cole, M. (2003) Vygotsky and context. Where did the connection come from and what difference does it make? Paper prepared for the biennial conferences of the International Society for Theoretical Psychology, Istanbul, Turkey, 22–7 June.
Lotman, J. (2005; first published 1984 in Russian) On the semiosphere (trans. W. Clark), *Sign Systems Studies*, 33(1): 205–29.

Stables, A. (2006) *Living and Learning as Semiotic Engagement.* New York: Mellen.
Stables, A. (2016) Education as process semiotics: Towards a new model of semiotics for teaching and learning. *Semiotica*, 212: 45–58.
Stables, A., Nöth, W., Olteanu, A., Pesce, S. and Pikkarainen, E. (2018) *Semiotic Theory of Learning: New Perspectives in Philosophy of Education.* London: Routledge.
Vygotsky, L.S. (1987) *The Collected Works of L.S. Vygotsky. Vol. 1: Problems of General Psychology, Including the Volume Thinking and Speech*, ed. R.W. Rieber and A.S. Carton, trans. N. Minick. New York: Plenum Press.
Wertsch, J.V., del Rio, P. and Alvarez, A. (1995) Sociocultural studies: History, action and mediation, in J.V. Wertsch, P. del Rio and A. Alvarez (eds.), *Sociocultural Studies of Mind.* New York: Cambridge University Press, pp. 1–34.

Credit: Adam Mørk

2
SCHOOL DESIGN
What do we know?

The overall driver of the project was the question as to whether design matters. Does design make any difference to practice? The answer is not a simple yes or no, rather it calls for a consideration of the relationship between the two. In the, thankfully, distant past, psychology was preoccupied with the nature – nurture debate and pursued forms of analysis that assumed a linear additive relationship rather than seeking and developing more sophisticated accounts of the interrelationships between the two.

The relationship between design and practice has a similar history with suggestions that design alone can change behaviour locked in conflict with those that suggest that it has little or no impact. Neither argument has developed a sophisticated model of the relationship between the two. There has been recognition of the complex nature of the influences that are brought to bear on design and on the nature of the knowledge that is needed for design to 'work'.

> The struggles to agree upon what counts as design knowledge and its cultural identity can therefore be perceived as affecting and being affected by a complex system involving economy, production, social significance, consumption, use of objects, and so on.
>
> *(Carvalho and Dong, 2006, p. 484)*

One of the major challenges of the *Design Matters?* project was to develop an understanding of the Design – Practice relationship. This required an in-depth engagement with the relevant literature and several cycles of analysis, portrayal and interrogation of the data. The contradictions and dilemmas that arose in this process became the driving force of our thinking and the development of the arguments that we present in this book.

The research examined school designs and the pedagogic practices that were witnessed in them. The project collected students', teachers' and parents' perceptions of their school environment over two years after occupation of new designs from the Building Schools for the Future and the Academies Programme in England. The Building Schools for the Future initiative gave rise to designs that aimed to provide inspiring learning environments and exceptional community assets over an extended period. The intention was to ensure that 'all young people are being taught in buildings that can enhance their learning and provide the facilities that they and their teachers need to reach their full potential'.

8 School design: what do we know?

BSF, launched in 2004, was the government-sponsored building programme of new secondary schools in England in place in the first decade of the twenty-first century. The programme was overseen by Partnerships for Schools (PfS), a non-departmental public body formed through a collaboration between the then Department for Children, Schools and Families (DCSF), Partnerships UK and private sector partners. There were 15 waves of funding activity starting with 14 English local education authorities expanding to 96 by 2009. This was a very complex initiative that attracted both praise and significant criticism.

The initiative involved the decentralisation of funds to local education partnerships (LEPs) who were commissioned to build and improve secondary school buildings as well as co-ordinate and oversee the educational transformation and community regeneration that was envisaged (Figure 2.1).

> The aim is not just to replace crumbling schools with new ones, but to transform the way we learn. This represents a break with the old way of doing things and should change the whole idea of 'school', from a physical place where children are simply taught to one where a community of individuals can share learning experiences and activities.
>
> *(CABE, 2006: 1)*

Aspirations for the outcomes of BSF were couched in terms of collaboration between schools, the development of new forms of infrastructure, new models of school organisation, an enhanced teaching

FIGURE 2.1 BSF case study school

FIGURE 2.2 Wider consultation

force, new patterns of distributed leadership, personalised approaches to teaching and learning involving significant and novel use of ICT and new forms of central governance (Hargreaves, 2003).

These new schools were spoken of as 'new cathedrals of learning' that were to be designed through high levels of consultation with key interest groups including parents and children (Figure 2.2).

> The design process was to involve 'proper consultation with the staff and pupils of the school and the wider community' (DfES, 2002: 63) in order that 'authorities and schools will be able to make visionary changes and enable teaching and learning to be transformed' (DfES, 2003a: 7).

The term 'personalisation' was a common feature of many policy documents and although it was linked to a myriad of meanings, it generally became associated with shifts in modes of control over learning with students taking more responsibility for the selection, sequencing and pacing of their work in school. The personalised approach was to be made feasible through access to new technologies and the availability of a mixed economy of open and flexible spaces. The argument promoted in favour of this significant investment was couched in terms of transformation of learning and teaching along with enhanced participation and community involvement and engagement. Environmental sustainability was a major consideration especially with respect to energy usage.

Considerable emphasis was also placed on the need for new approaches to school leadership:

> Our determination is to ensure that every Head is able to do more than run a stable school. Transformation requires leadership which: Can frame a clear vision that engages the school community;

Can motivate and inspire; Pursues change in a consistent and disciplined way; and Understands and leads the professional business of teaching. To achieve their full potential, teachers need to work in a school that is creative, enabling and flexible. And the biggest influence is the Head . . . Heads must be free to remodel school staffing, the organisation of the school day, school week and school year and be imaginative in the use of school space – opening up opportunities for learning in the community, engaging with business and developing vocational studies.

(DfES, 2002 p. 26)

However, as Kraftl (2012) points out, there is some doubt as to whether this radical vision of restructuring was realised in the realities of practice in schools and communities:

BSF connected with the promise of three further discourses: school, community and architectural practice. It anticipated that new school buildings would instil transformative change – modernising English schooling, combating social exclusion and leaving an architectural 'legacy'. However, it is argued that BSF constituted an allegorical utopia: while suggesting a 'radical' vision for schooling and society, its ultimate effect was to preserve a conventional (neo-liberal) model of schooling.

(p. 847)

More recently the subject of design quality in schools has come to the fore with government pronouncements on the wastage of money on architectural fees and what has been referred to as over-indulgent design within the BSF and Academies Programme. The architectural profession has responded that they had been asked to produce higher-quality environments particularly in terms of the acoustic environment, the quality of daylighting and higher-quality ventilation, the provision of ICT, and the reduction in energy costs (RIBA, 2017; Figure 2.3).

FIGURE 2.3 BSF case study school

Some buildings may prove extremely good value for money in terms of their impact on the educational achievements of their pupils; others may not (James, 2011).

The purpose of *Design Matters?* was to assess both qualitatively and, as far as possible, quantitatively what that impact of design has been on the sample of schools in our project. The policy environment at that time was one in which capital investment was made in order to secure radical change in the practices of schooling. Teaching, learning, management and community participation and engagement were to be transformed as new schools were designed and built to meet the envisaged needs of the twenty-first century. More recently, policy on the role of design in rebuilding the schools estate in England has been though another major change as attempts are made to achieve good value and efficiency in times of austerity. On 5 July 2010 the then Secretary of State for Education, Michael Gove, announced that the BSF programme was to be scrapped. The Priority School Building Programme (PSBP) was established in 2011 to rebuild, or meet the condition needs of those schools that remained in the worst condition. The PSBP is seeking to reduce school building costs by approximately a third in comparison with those incurred during BSF. Project time has also been reduced from 24–36 months to 12 months in order to drive efficiency. This involves limiting consultation with school communities and multiple stakeholders to an initial six-week period. The Education Funding Agency (EFA) (now the ESFA) has produced Control Options in order to demonstrate how their Baseline Designs should be applied in practice (Figure 2.4).

> Good quality education does not necessarily need sparkling, architect-designed buildings . . . Throughout its life [BSF] has been characterised by massive overspends, tragic delays, botched construction projects and needless bureaucracy.
>
> *(Gove, 2010, cited in Kraftl, 2012, p. 866)*

FIGURE 2.4 Control options

There is very limited research on the performance of the BSF designs and their fitness for the purposes of occupiers. Importantly Parnell et al. (2008) point to the necessity of a clear pedagogic vision for transformation and effective practices of communication between key stakeholders if the outcomes of new-build projects are to be fit for the purposes of those who occupy and work in the buildings once they are complete:

> Developing the educational vision with the whole school and an educational philosophy which can be communicated to the designer/architect ensures that the school has a strong pedagogical basis stemming from the people within, rather than being imposed by those without, or by the form of a building. Representatives from local authorities who had initiated engaging stakeholders in the process were clearly committed to the opportunities to rethinking learning with the new school buildings. This is reflected in the names for new schools, such as 'new learning centres' which are replacing secondary schools in Knowsley and in Birmingham the BSF projects are part of the Council's Transforming Education Programme. Similarly, the contributing contractor showed a high level of awareness of and commitment to the goal of educational transformation.
>
> *(p. 216)*

Current policy has placed a significant emphasis on the growth and development of the UK design and construction industry as one that involves multiple stakeholders often with different priorities and perspectives (BIS, 2013). Sadly, a recent survey by The Key (2015) suggested that 35 per cent of headteachers felt their school buildings were not fit for purpose. Parnell et al. (2008) point to the potential challenges that differences in motivation can cause and give rise to the kind of dissatisfaction:

> This study has revealed a complex process of design with potential for collaborative working between schools, architects, contractors and local authorities. Although the core goal – the realisation of a new or remodelled place to learn – is shared, the process of realising this place provides participants with different opportunities. Each stakeholder inevitably prioritises outcomes based on their differing motivations, which can potentially create tensions.
>
> *(p. 222)*

These issues have been noted by many of the critiques of the working practices of the design and construction industry. In 1994 the Latham Report (Constructing the Team; Latham, 1994), investigated the perceived problems with the construction industry, describing it as 'ineffective', 'adversarial', 'fragmented' and 'incapable of delivering for its customers', and proposing that there should be greater partnering and teamwork. This message was reinforced by the 1998 Egan Report (Rethinking Construction) and the Government Construction Strategy in 2011, which made similar, somewhat damning assessments, of the industry. Importantly, Morrell (2015) has put forward argument for the centrality of collaboration as a driving force for change in the construction industry. What is arguably a need for a cultural shift within the industry still seems to be an aspiration sometimes cast in rhetoric which is yet to be realised in practice:

> Shifting working and organisational cultures similarly requires time and might even be too much to ask. However, this could be what is required to engender a truly collaborative working process which gives all participants a voice. Where appropriate, architectural practices and building contractors, which seek to specialise in collaborative school design, might need to work towards this cultural shift. Similarly, schools can begin to embrace student (and teacher) participation in the wider culture of the school.
>
> *(Parnell et al., 2008, p. 223)*

These enduring challenges may well require new ways of conceptualising collaboration which recognise the formation of constantly changing combinations of people and resources across agencies, and their distribution over space and time. Discussions with designers and contractors suggest that such work demands changes both in inter-professional attitudes and practice and relationships with clients. This form of work has been characterised as co-configuration: a form of work orientated towards dynamic, reciprocal relationships between providers and clients (cf. Victor and Boynton, 1998).

This form of work is witnessed in attempts that have been made to personalise services for children and young people. There is a well-rehearsed account of the importance of dialogue as a means of improving performance in learning and decision making. It is also clear that when young people are involved in the decisions about the education and social care which is made available for them, they become more highly motivated and the provision is more likely to be effective. A crucial part of this involvement would seem to be self-assessment. Practices of self-assessment are not only important during the years of schooling. The personalisation of public services, promoted as the next step in the modernisation of the welfare state (Leadbeater, 2004), positions clients as co-producers of services with a central role in their design. Personalisation requires citizens who are capable of participating in dialogues about their needs and desires as well as their own interpretations of their current situation. Just as Black and Wiliam (1998) argued that teachers and pupils should be prepared for self-assessment in schools, so the personalisation agenda brings questions about the ways in which the most vulnerable are to be prepared for participating in dialogues about their futures. The principles of participation, personalisation and capability pose challenges to our knowledge base when we consider their application to those who experience difficulty in learning and in communication.

In practices of co-configuration there is a need to go beyond conventional teamwork or networking to the practice of 'knotworking' (Engeström, 1999). Engeström argues that knotworking is a rapidly changing, distributed and partially improvised orchestration of collaborative performance which takes place between otherwise loosely connected actors and their work systems to support clients. In knotworking various forms of tying and untying of otherwise separate threads of activity takes place. Co-configuration in responsive and collaborating services requires flexible knotworking in which no single actor has the sole, fixed responsibility and control. It requires participants to have a disposition to recognise and engage with the expertise distributed across rapidly changing work places. As Engeström and Middleton (1996) suggest, expertise in such contexts is best understood as the collaborative and discursive construction of tasks, solutions, visions, breakdowns and innovations. A precondition of successful co-configuration work is dialogue in which the parties rely on real-time feedback information on their activity. The interpretation, negotiation and synthesising of such information between the parties requires new, dialogical and reflective knowledge tools as well as new, collaboratively constructed functional rules and infrastructures (Engeström and Ahonen, 2001).

These two aspects of learning are evident in organisational, interactional and discursive practice in knotworking in inter-professional working. Learning in co-configuration settings is typically distributed over long, discontinuous periods of time. It is accomplished in and between multiple loosely interconnected activity systems and organisations and represents different traditions, domains of expertise and social languages. In short, we see learning as being able to interpret our worlds in increasingly complex ways and being able to respond to those interpretations. How we respond as professionals very much depends on whether the workplace allows the responses that are necessary. We therefore argue that individual learning cannot be separated from organisational learning.

This form of professional action is called for in current policy (BIS, 2013). For example:

> We believe that, as a nation, we need to recognise the power of place and to be much more ambitious when planning, designing, constructing and maintaining our built environment. Failure to do

so will result in significant long-term costs. We now set out some of the important measures that need to be taken to achieve this aim.

(House of Lords Select Committee on National Policy for the Built Environment, 2016, para 64, p. 92)

The relatively limited research on the working practices of the school construction industry suggests a better understanding of current practices is required before significant innovation and improvement can take place.

There are therefore a number of under-researched areas with respect to school design, construction and occupation. Rather than engage with all the literature concerned with school design, we will present an overview of the arguments concerned with design, construction and occupation of new schools which seek to meet the demands of new and emergent forms of pedagogic practice. This overview will start with a consideration of the relationship between design and practice and then move to an examination of the arguments concerning the likely futures for design and practice. This will be followed by a discussion of the relationship between the processes of design and construction. The chapter will conclude with an overview of discussions concerning post-occupancy evaluation (POE) and the ways in which we learn about the outcomes of design and build projects.

The relationship between design and practice

As will become clear in subsequent chapters of this book, we studied new-build schools that policymakers thought would transform practice. In this section we will review the evidence concerning the nature and the effects of the relationship between design and practice.

Earthman (2004) concludes that while inadequate school buildings cause 'health problems, lower student morale and contribute to poor student performance', he is not convinced that school buildings need necessarily be any more than adequate, although the notion of adequacy fails to find a satisfactory definition. Rutter, who found no relationship between physical environmental factors and a range of learning and behavioural outcomes, comments that, 'It was entirely possible for schools to obtain good outcomes in spite of initially rather unpromising and unprepossessing school premises' (Rutter, 1979, p. 178).

As shown in Woolner and Thomas (2016), compelling research demonstrates that different physical configurations facilitate some pedagogical approaches while hindering others. Specifically, traditional classrooms are more likely to promote teacher-centred practices and suppress student collaboration (Figure 2.5) (Sigurðardóttir and Hjartson, 2011), while open-plan schools foster student involvement, teacher cooperation, collaboration and, often, team-teaching (Figure 2.6) (Gislason, 2010, 2015; Saltmarsh et al., 2015).

> There is clear evidence that extremes of environmental elements (for example, poor ventilation or excessive noise) have negative effects on students and teachers and that improving these elements has significant benefits. However, once school environments come up to minimum standards, the evidence of effect is less clear-cut. Our evaluation suggests that the nature of the improvements made in schools may have less to do with the specific element chosen for change than with how the process of change is managed. There appears to be a strong link between effective engagement with staff, students and other users of school buildings and the success of environmental change in having an impact on behaviour, well-being or attainment. The ownership of innovation, in contrast to the externally imposed solution, appears to tap directly into motivational aspects which are key

FIGURE 2.5 Traditional classroom

FIGURE 2.6 Teaching and learning environment with a mixed economy of space

factors in maximising the impact of change. Changing the environment is 'worth doing' if it is done as a design process.

(Higgins et al., 2005, p. 6)

A recent review conducted by OECD (2013) sought to identify how 'investments in the physical learning environment – that is "the physical spaces (including formal and informal spaces) in which learners, teachers, content, equipment and technologies interact" – can translate into improved cognitive and non-cognitive outcomes' (p. 1). In order to do this they explored the ways in which spatiality, connectivity and temporality mediate pedagogical and other relationships that can improve student learning. The emphasis here on mediation is important. It suggests a very different mechanism is at play than one of determination. They recognised that empirical evidence was far from extensive and agreed with Woolner et al. that:

> the research indicates that there is an overall lack of empirical evidence about the impact of individual elements of the physical environment which might inform school design at a practice level to support student achievement.

(2007, p. 47)

More recently however, Barrett et al. (2015) have suggested that differences in the physical characteristics of primary school classrooms explain 16 per cent of the variation in learning progress. Their claim is that theirs is the first time that clear evidence of the effect on users of the overall design of the physical learning space has been isolated in real-life situations. Their findings point to a classroom rather a whole school design effect:

> Surprisingly, whole-school factors (eg size, navigation routes, specialist facilities, play facilities) do not seem to be anywhere near as important as the design of the individual classrooms. This point is reinforced by clear evidence that it is quite typical to have a mix of more and less effective classrooms in the same school. The message is that, first and foremost, each classroom has to be well designed.

(Barrett et al., 2015, p. 3)

However, the purpose of their sampling strategy was 'to achieve as much variety in the sample as possible, while still focusing on mainstream schools where there is some consistency in the broad context' (p. 10) rather than to model and sample the whole school designs.

The nature of evidence in such matters is a topic that is not without its controversies. Goldacre (2013) has made a plea for the development of evidence-based practice in education. His argument cites the importance of randomised controlled trials (RCTs) in practices such as medicine. An important adjunct to this claim is the one advanced by Cartwright and Hardie (2012) who argue that evidence alone is not enough and that policymakers must be able to access relevant evidence if their policy is to work.

There is a pressing need to consider the nature of the assumed causal mechanism that underlies studies of the relationship between design and practice and the extent to which such mechanisms take account of the context in which evidence may be relevant:

> Evidence on the impact of previous building programmes needs to be urgently assembled and employed in order to help policymakers develop both initiatives and instruments that are genuinely capable of delivery. This will require the involvement of researchers, practitioners, building

managers and others in the development of a robust body of accessible knowledge and an evidence enabled predictive capacity. Better and more effective policy is capable of achieving a great deal more than any number of good individual projects, but to truly realise this potential, a move is required to create a more policy literate and committed built environment sector capable of well-informed design at both project and policy levels.

(Foxell and Cooper, 2015, p. 405)

The OECD review (OECD, 2014) argues that space '*shapes*' social relations and practices in schools and communities. They further suggest that social practices, formal instruction and informal social interactions change the nature, use and experience of space and that in turn varies for individuals and groups according to gender, ethnicity, race, religion and disability (p. 10).

There have been various attempts to evidence these effects using both qualitative (e.g. Heppell et al., 2004) and quantitative (e.g. Higgins et al., 2005) research designs.

A significant proportion of the research on school design focuses on environmental issues such as acoustics, lighting and temperature and only offers a quantitative analysis (e.g. Hygge, 2003; Galasiu and Veitch, 2006; Winterbottom and Wilkins, 2009; Shaughnessy et al., 2006). Recent empirical work continues to suggest a relationship between the built environment and learning potential (e.g. Barrett et al., 2013). Furthermore, while there is a body of research which explores the relationship between buildings and pedagogies (e.g. Burke, 2010; Burke and Grosvenor, 2003, 2008), to our knowledge, there is a gap in research which explores the relationship of an educational vision to the creation, use and maintenance of physical space and, more importantly, how this relationship impacts on perceptions of educational experiences.

The selective use of what counts as evidence continues to be a feature of the policymaking landscape. The relatively recent major policy change with regard to the degree of central control over school design was made amid claims of 'no firm evidence' of a relationship between school renovation and improved results (Vasagar, 2012) despite the fact that, as Mahony and Hextall (2013, p. 862) point out that there is a 'a growing body of research on the educational effects of newly designed schools (e.g. Higgins et al., 2005; Woolner et al., 2007; PriceWaterhouseCoopers LLP, 2007, 2008, 2010).

The argument that we will advance in this book is that in order to understand the place of design in the performance of a school, it is vital to understand the way in which design and practice are brought into relation with each other. This applies as much to the social relations in the practices of design and construction as it does to occupation. The three quotes provided below give some insight into the complexity of these relationships:

The task ahead is to see all social phenomena as emplaced, as being constituted in part through location, material form, and their imaginings (Appadurai, 1996). Put more tractably, place stands in a recursive relation to other social and cultural entities: places are made through human practices and institutions even as they help to make those practices and institutions (Giddens 1984). Place mediates social life; it is something more than just another independent variable.

(Gieryn, 2000, p. 467)

Buildings (factories or laboratories) do as much to structure social relations by concealing as by revealing, and therein lies their distinctive force for structuring social relations and practices. Once completed, buildings hide the many possibilities that did not get built, as they bury the interests, the politics, and power that shaped the one design that did.

(Gieryn, 2002, p. 39)

> Buildings may be comparable to other artefacts in that they assemble elements into a physical object with a certain form; but they are incomparable in that they also create and order empty volumes of space resulting from that object into a pattern. It is this ordering of space that is the purpose of building, not the physical space itself . . . the ordering of space in buildings is really about the ordering of relations between people.
>
> *(Hillier and Hanson, 1984, pp. 1–2)*

These quotes come close to the position that has influenced the thinking and some of the challenges that were involved in the research processes involved in *Design Matters?* This complex dialectical view of the relationships between buildings, human action including management, social organisations and social structures informs the way we studied the schools, their designers, constructors and occupiers.

Alongside the debates about the design of new schools during the BSF period there was a wide-ranging discussion about the nature of pedagogic practices that were deemed to be fit for the future. Issues such as personalised learning, individual pathway planning, team teaching, inquiry approaches, student teamwork, problem solving, rich tasks and community-based service learning were raised often in the context of the rapid development of new technologies for learning. The need for adaptable and flexible spaces that could support a mixed economy of pedagogic approaches mediated by these new technologies was clearly recognised by senior politicians of the day:

> We need to look at ways of designing inspiring buildings that can adapt to educational and technological change. ICT can give schools the option of teaching children as individuals, in small groups and in large groups, and can provide electronic links to other schools and facilities in this country and abroad. That will not happen if we do not design spaces in schools that are flexible and will facilitate various patterns of group working. Flexibility is key, because whatever visions of education we design our buildings around, we can be sure that they will need to perform in a very different way in a few years' time.
>
> *(David Milliband in DfES, 2003c, p. 3)*

One County Council (CC) in our sample commissioned Microsoft Consulting Services to prepare a White Paper in support of the development of its BSF vision with a particular focus on the potential that ICT has to enable system wide transformation of its education service (Figure 2.7). They set an agenda that envisioned a radical transformation the practices of education:

> Education, together with sustainable economic development, can provide the catalyst to inspire learners with the skills and knowledge to become confident, self reliant, healthy, collaborative and responsible citizens who are economically active and able to participate in a democratic society. BSF must therefore focus on the individual and their potential to achieve, taking full account of the way in which young people increasingly lead their lives in a digital and connected world. In this sense BSF is not just about buildings. It is not just about transforming learning. BSF has to be about realising the potential of every young person and improving their life chances in a new knowledge economy.
>
> *(Microsoft Consulting Services, 2005, p. 3)*

These radical and highly optimistic claims were set alongside aspirations to connect regeneration and renewal agendas with the lifelong learning agenda.

FIGURE 2.7 Example of open-plan space with ICT

The government of the day argued we need a more strategic approach to the future development of ICT in education, skills and children's services. They believed that this would:

- Transform teaching, learning and help to improve outcomes for children and young people . . .
- Engage 'hard to reach' learners with special needs, support more motivating ways of learning, and more choice about how and where to learn
- Build an open accessible system, with more information and services online for parents and carers, children, young people, adult learners and employers; and more cross-organisation collaboration to improve personalised support and choice
- Achieve greater efficiency and effectiveness, with on-line research, access to shared ideas and . . . plans, improved systems and processes, shared procurement and easier administration.

(DfES, 2005, p. 4)

Research conducted at about the same time indicated that the same principles of flexibility and personalisation were important components of inclusive practice. Dyson et al. (2004) concluded from their government-funded study of inclusion and pupil achievement:

Highly-inclusive and high-performing schools adopt a model of provision based on flexibility of grouping, customisation of provision to individual circumstances and careful individual monitoring, alongside population wide strategies for raising attainment.

(p. 1)

They attribute this approach to a school-level commitment, which in their earlier work they suggest can be promoted by SENCOs. In turn, this form of practice is commensurate with forms of distributed leadership that have been advocated elsewhere in the development of inclusive schools (Mayrowetz and Weinstein, 1999).

Effective use of ICT in education requires that teachers are able to change their practices to be more student-centred and to give over control, and that students are capable of self-regulated learning in the classroom and online (Luckin, 2010).

Interestingly, the role of the teacher in shifting students from more contextual, everyday knowledge to more abstract, academic could have been predicted by the theories of Vygotsky (1987) and more recently by the work of Young and Muller (2010), who argue for the importance of recognising boundaries between everyday and school knowledge. However, this does contrast with the rhetoric around schools of the future with its suggestion that the use of digital technologies in schools will lead to an erosion of boundaries between home and school (Sutherland and Fischer, 2014. p. 3). There are echoes here of debates that took place in the 1970s concerning the development of project-based curriculum projects which focused heavily on the local everyday experience of young people at the expense of the development of a more 'scientific' understanding and conceptual development. In some respects it is as if the collective memory on such matters has been eroded. We have found very little by way of reference to the open-plan schools experiments of the same period within the debates about the design of schools with a mixed economy of space (Watts, 1977; Cooper, 1985).

These understandings of a mixed economy of space posit the development of school designs driven by pedagogies based on personalised learning. In these circumstances in which considerable use may be made of technologies with students moving from large open flexible areas to smaller spaces for learning (Figure 2.8).

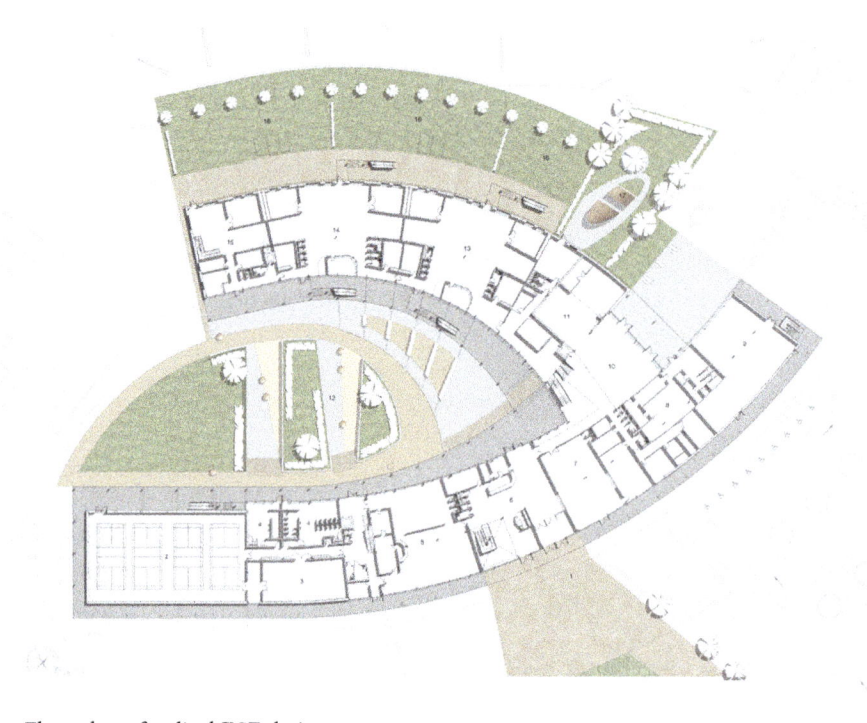

FIGURE 2.8 Floorplan of radical BSF design

Credit: HKS Architects

In the *Design Matters?* study we were keen to look at student movement through the spaces of these new school designs to try to ascertain whether the practices that were envisioned were actually taking pace. Certainly in the 1970s open-plan designs Cooper (1985) observed that the flexibility in use of space that was intended was curtailed by both physical (in the form of ad hoc barriers) and discursive (in the form of teacher directions) constraints.

OECD (2013) point to the professional learning, leadership and systemic redesign of some aspects of proposed changes in school design. However, for the assumptions to be made manifest in practice, there is a pressing need to manage and lead the way forward into new ways teaching of learning. Put simply, new forms of practice have to be learned, they will not be enforced through design (Blackmore et al., 2011). One way of thinking about a new school design is that it constitutes a new tool for working on educational challenges. Some of the suggestions that were advanced for the use of the BSF designs represent fundamental changes in the work of teachers. There are considerable professional challenges for teachers who have worked in a traditional 'single cell' class with 30 students when they move into a school designed for large groups of students who, in the pursuit of personalised learning agendas, may be moving from large open spaces to smaller closed spaces either individually or in small groups that may be formed and reformed. Similarly managing these new and complex systems bring considerable challenges for school management.

Another feature of the processes of new school design may also be understood in terms of learning. This is concerned with how the different stakeholders and participants involved in the process learn from each other about the different perspectives and priorities that are held for the future school. Among others, Pam Woolner and her colleagues have conducted research on ways of involving students in processes of consultation. Others have commented on the importance of general processes of consultation as well as the involvement of teachers (Fisher, 2005; Higgins et al., 2005; Johnson et al., 2000; Morgan, 2000; Radcliffe et al., 2008). Lack of effective consultation has been posited as the root cause of design failure:

> When environmental change fails to change teachers' and learners' behaviour, this may well be because issues of communication have not been addressed and systems and processes have, therefore, failed to adapt to meet the change in the environment.
>
> *(Higgins et al., 2005, p. 37)*

In the *Design Matters?* project we studied the processes of consultation and communication throughout the practices of design and construction

Much of the research points to resistance to change as a key issue in the use of new designs which often involves little change in pedagogic practice. Woolner and colleagues point to the need for consultation and active participation on the part of end-users:

> School premises make a difference to learning, but it is important to understand the relationship between setting and educational activities. Physical space has been found to entrench practice, making it harder to reflect and make changes. Yet changes made to the physical environment may not lead to changes in teaching or learning. This may be understood theoretically in terms of levels of participation, and many school design practitioners advocate active participation of school communities in processes of change.
>
> *(2011, p. 1)*

However, there is always the possibility that consultation may act as a rhetorical device which generates data which are either systematically ignored or discounted by professionals involved in the processes of design. Den Besten et al. (2008) used interview data gathered in local authorities to demonstrate how the priorities of budgets, personalities and local contingencies subverted the intentions of consultation.

Research also suggests that teachers in some curriculum areas may be more open to change than others. While new buildings may make end-users feel better about their school they may not change their practice:

> The analysis developed in this article shows that the renovation of school spaces resulted in improved conditions, which, according to the perspective of school boards, teachers and pupils, increased pride in and responsibility for the school. But teaching practices in schools remain set in a traditional mould, with the prevalence of teachers' lectures, while pupil-centred and environment-oriented activities are scarcer. Although this is the most dominant behaviour among teachers, it is possible to identify some variations when analysing the schools individually, notably in schools with a prevalence of vocational training, in which pupil-centred activities are more common, and in art schools, where environment-oriented activities prevail.
>
> *(Veloso et al., 2014, p. 418)*

As Woolner et al. (2012) note, resistance to changes in school practices may be because the changes have been imposed, rather than a result of teacher and pupil awareness of the reasons for those changes. That is why it is important to outline participation models in the design process, not only for consultation purposes, but also as part of a comprehensive discussion of learning models and practices, in order to attain a shared understanding among staff and students of what learning is, or could be (Woolner et al., 2012).

In the *Design Matters?* project we were keen to identify the forms of professional development that were invoked in order that teaching professionals may learn to make best use their newly built schools.

Current views on the outcomes of the BSF programme are not overwhelming supportive. For example, the Commission for Architecture and the Built Environment (CABE), which audited 52 of the 104 schools refurbished between 2001 and 2006, found 31 per cent of the design quality of the school buildings it inspected 'poor', 21 per cent 'mediocre', 29 per cent 'partially good', 15 per cent 'good' and 4 per cent 'excellent' after initial occupation but the how the buildings support pedagogic practice in the longer term is still poorly understood.

The most influential source of criticism was the government commissioned James Report, published in 2011. Among many other criticisms the report bemoans the lack of attention to learning from experience from within the programme:

> The design and procurement process for the Building Schools for the Future programme (and other strategic programmes) was not designed to create either high and consistent quality or low cost. Procurement starts with a sum of money rather than with a specification, designs are far too bespoke, and there is no evidence of an effective way of learning from mistakes (or successes).
>
> *(James, 2011, p. 5)*

This finding of lack of learning about design processes in practice paved the way for a policy move towards standardisation and a higher degree of central control which now features in the Priority School Building Programme. Policy on the role of design in rebuilding the schools estate in England is at a crossroads as attempts are made to achieve good value and efficiency in times of austerity:

> New buildings should be based on a clear set of standardised drawings and specifications that will incorporate the latest thinking on educational requirements and the bulk of regulatory needs. This

will allow for continuous learning to improve quality and reduce cost. Currently the bulk of new schools are designed from scratch with significant negative consequences on time, cost and quality.

(James, 2011, p. 6)

In the *Design Matters?* project we were very keen to understand the influences that had been brought to bear on design processes and how professionals had learned about what mattered in effective and efficient school design. We were asking the same sort of question that Gieryn posed: 'How do places come to be the way they are, and how do places matter for social practices and historical change' (2000, p. 463).

On building performance: tackling the divide between what is promised by the industry and what is delivered – the performance gap, taking responsibility for the whole-life of projects, developing common metrics; and committing to measurement and evaluation, and the dissemination of findings. There can hardly be a cause more deserving of cross-industry collective contemplation and action than the constant failure of its product.

(Morrell, 2015, p. 8)

Limited research on the working practices of the construction industry suggests a better understanding of current practices is required before significant innovation and improvement can take place. This form of professional action is called for in current policy (BIS, 2013).

From 2016 all central government construction capital projects will be required to utilise two tools to address these issues: the implementation of building information modelling (BIM) and Government Soft Landings (GSL). BIM allows buildings to be conceived collaboratively and tested virtually, before they are built and operated for real (Eastman et al., 2011). Although BIM technologies may enable tighter links between project participants, in practice they will require a significant change in culture (Kerosuo, 2015).

The collaborative nature of BIM means that it is most effective when a procurement route is adopted that itself facilitates collaboration; such as the early involvement of suppliers and contractors. Small-scale studies show, however, that the problems of working together in design and construction teams have not decreased with the adoption of BIM (Volk et al. 2014). Indeed Harty and Whyte (2010) suggest that practitioners develop 'work arounds' which make it appear that BIM is being used whereas in practice traditional methods are deployed and obscured from external view. Such rule-bending may be taken as a sign that organisations are not keeping pace with the demands of practices, in line with Barley and Kunda's (2001) observation that when the nature of work changes, organisational structures either adapt or risk becoming misaligned with the activities they organise. 'Soft Landings' is the British Services Research and Information Association-led initiative that aims to solve the performance gap between design intentions and operational outcomes (Tuohy and Murphy, 2015). It aims to increase designer and constructor involvement before and after handover, and points the 'supply side' to more involvement with users and a careful assessment of building performance in use. However, very limited research has been carried out on Soft Landings in use and its effectiveness in enhancing collaborative practices.

As Porter (2016) notes, schools are particularly complex environments with a wide range of end-users with rapidly changing demands on the performance of the designs. (Figure 2.9)

Arguably, therefore, those who are least resilient to the effects of buildings that are not fit for purpose are those who find schooling, in general, challenging. Additionally there is a lingering concern about the extent to which a design which is fit for purpose for those regarded as 'mainstream' is regarded as fit for purpose for those who have additional needs. BB 105 urges designers 'to think about children's SEN and disabilities right from the start, placing them at the heart of all stages of the design process' (BB 102, 2012). These issues are not addressed in the current guidance on access and inclusion for the PSBP. Children may

FIGURE 2.9 Special needs school with central atrium space
Credit: Hunters South Architects

well have competing needs (Clark, 2002) demanding innovative solutions rather than a formulaic response to technical specifications (Gathorne-Hardy, 2001) For example, children that are immobile require different temperatures to those who are highly active (Clark, 2002). Children with visual impairments often rely on tactile and auditory cues (including echoes) to navigate their way around environments, all aspects that children with Autistic Spectrum Disorders can find highly distracting and disruptive, which consequently impact negatively on their learning (Tufvesson and Tufvesson, 2009). Specialist spaces provide partial answers if children are effectively excluded from participating in valued activities and social contact with their peers (Hodson et al., 2005). Additional and competing needs provide a further layer of complexity in negotiation around the decision-making process (Porter, 2016).

Post-occupancy evaluation: what does it measure?

Post-occupancy evaluation (POE) research is typically aimed at measuring the performance of aspects such as lighting, temperature and acoustics (e.g. Hygge, 2003; Galasiu and Veithch, 2006; Winterbottom and Wilkins, 2009; Shaughnessy et al., 2006) with the intention of informing architects, educators and policymakers on what type of building creates optimal learning conditions in relation to such factors. Environmental POEs typically only contain quantitative data and tell us little about how different environmental factors interact with users through time (Tse et al., 2014).

The work of Burke and Grosvenor (2008) does provide an insight into how school buildings can mediate students' perceptions of who they are, what they should think about the world, expected ways of behaving and feelings of well-being, thus developing Churchill's claim that 'First we shape our buildings; thereafter they shape us.' However, it is this 'shaping' process which is less frequently researched (e.g. OECD, 2014) even though 'direct and psychosocial influences are apparent on all school users' (ibid.). The limited research in this area also fails to offer a systematic analysis of how space is shaped (Woolner et al., 2007) and, to our knowledge, there is no research which explores how space is shaped from the inception of a design idea through to occupation and the mediating effects of this space on end-users.

Post-occupancy evaluation (POE) allows benchmarks and patterns to be established that can be captured and used to inform future design. POE will also inform the decision of how best to invest future capital expenditure for maximum return. The British Council for School Environments (BCSE) conducted a review of post-occupancy evaluation methods, concluding that a POE framework should be set up to inform school design on a national and even international level (BCSE, 2009).

However, such a framework would be very unlikely to be as broad in scope as the present proposal. Current architectural research on school environments tends to focus on environmental performance in terms of factors such as natural daylight, air quality, temperature and noise level. Woolner et al. (2007) warn that such research in isolation can lead to confusing, and often contradictory, conclusions; however, they also acknowledge that inadequate temperature control, lighting, air quality and acoustics have detrimental effects on concentration, mood, well-being, attendance and, ultimately, attainment.

Such research is therefore relevant, but educationally limited. By contrast, Moos (1979) argued that the learning environment is best understood as resulting from a complex interaction of social, cultural, organisational and physical factors. Benito (2003) directed attention to the meanings of school design and the cultural function that is assigned to schools. From this perspective school architecture should be open to a form of analysis which takes account of educational discourses and practices, and actors' social norms. Cooper (1981) examined the conflicts that arose from differences in pedagogic orientations expressed in school design from those adhered to by most teachers. More recently Leiringer and Cardellino (2011) have argued how important it is to find a balance between good design, commercial realities and educational approaches. This points to the need to understand the ways in which the philosophies and discourses of design and educational practice intersect at particular moments and over time. Burke and Grosvenor (2008) examined the history of the relationship between school design and educational philosophy/practice while Cooper (1985) has argued that school building may be regarded as the physical embodiment of the educational system and the changing philosophies which inspire it. Prosser (2007) directed attention to the ways in which teachers' and pupils' everyday behaviours shape and in turn are shaped by school culture which is manifested in part visually in the built environment. Cooper (1985) also noted the importance of non-teaching spaces, which are taken-for-granted yet deeply embedded in the teaching and learning behaviours of generations of teachers and pupils. Prosser (2007) argues that the design of schools reflect both developments in educational philosophy, as aims are re-defined, and new physical standards and methods of construction. Given respective timescales, Dudek (2000) argues that building to match educational theory is implausible. Research on the participation of school users has focused on commissioning and design. Woolner et al. (2005) caution that the history of school building programmes is littered with supposedly innovatory design which subsequently becomes unfit for future purpose. They stress the importance of user engagement in defining and solving design problems, so that successful solutions come to be seen as flexible and adaptable to new learners and teachers, curriculum demands and challenges (Woolner et al. 2007: 64). This emphasis on user engagement in design processes is taken up by Clark (2010) who refined the 'Mosiac' approach (Clark and Moss, 2005) as a method for listening to and collecting data from children. Thus approaches to user involvement have been developed. The extent to which they are deployed in commercial POE is another matter.

Conclusion

We have not discussed all the literature that we have accessed in the course of *Design Matters?* We made decisions about relevance and have presented an account of what we take as the key issues. On the basis of this exercise we identified a number of key issues which we took forward into the main body of the research

The overriding issue is the relationship between design and practice. This calls for a thorough examination of both components. With respect to the processes of design and construction there is a need to understand the nature and effects of the consultations that took place including accounts of children's experience. There is also a need to understand the relationships between key stakeholders in the design and construction process and the extent to which learning from the perspective of others and from past experience was a part of professional practice. This will also require an understanding of the tensions and dilemmas that arose at particular stages and over the entire time span of the process.

In the practices of occupation we were minded to consider the extent to which educational practice was personalised and innovatory pedagogy was developed. The way in which space was used and the experience of using space, including differentials by factors such as gender, in that experience were also matters of importance. The use of space within lesson time as well at social times was considered a priority. Alongside this use of space issue is the use of ICT concern. The overall impact on well-being as well as academic achievement came out as a priority.

As far as management of the schools was concerned we were directed towards a consideration of management style and the extent to which the school was organised in such a way that professionals could learn to use the new building. Relationships with the local community were also considered as a valuable point of scrutiny. Lastly the financial arrangements and their impact on both design and practice appeared to be matters of some importance.

References

Abu-Lughod, J. (1968) The city is dead – Long live the city. In S.F. Fava (ed.), *Urbanism in World Perspective*. New York: Crowell, pp. 154–65.
Appadurai, A. (1996) *Modernity at Large: Cultural Dimensions of Globalisation*. Minneapolis: University of Minneapolis Press.
Barley, S. and Kunda, G. (2001) Bringing work back in, *Organisation Science*, 12(1): 76–95.
Barrett, P., Zhang, Y., Moffat, J. and Kobbacy, K. (2013) An holistic, multi-level analysis identifying the impact of classroom design on pupils' learning, *Building and Environment*, 59: 678–89.
Barrett, P., Zhang, Y., Davies, F. and Barrett, L. (2015) *Clever Classrooms: Summary Report of the HEAD Project*. Salford: University of Salford.
Benito, A.E. (2003) The school in the city: School architecture as discourse and as text, *Pedagogica Historica*, 39(1–2): 53–64.

Black, P. and Wiliam, D. (1998) Assessment and classroom learning, *Assessment in Education: Principles, Policy and Practice*, 5(1): 7–74.

Blackmore, J., Bateman, D., Cloonan, A., Dixon, M., Loughlin, J., O'Mara, J. and Senior, K. (2011) *Innovative Learning Environments Research Study*. East Melbourne, Vic.: Department of Education and Early Childhood Development. https://dro.deakin.edu.au/eserv/DU:30036968/blackmore-researchinto-2011.pdf

British Council for School Environments (BCSE) (2009) www.bcse.uk.net/menu.asp?id=50920.

Building Bulletin (BB) 102 (2012) Designing for disabled children and children with special educational needs: Guidance for mainstream and special schools. London: Department for Children, Schools and Families (DCSF).

Burke, C. (2010) About looking: Vision, transformation and the education of the eye in discourses of school renewal part and present, *British Educational Research Journal*, 36(1): 65–82.

Burke, C. and Grosvenor, I. (2003) *The School I'd Like. Children and Young People's Reflections on an Education for the 21st Century*. London: RoutledgeFalmer.

Burke, C. and Grosvenor, I. (2008) *School*. London: Reaktion Press. Cartwright, N. and Hardie, J. (2012) *Evidence-based policy: A practical guide to doing it better*. New York: Oxford University Press.

Carvalho, L. and Dong, A. (2006) Legitimating design: A sociology of knowledge account of the field. *Design Studies*, 30(5): 483–502.

Clark, A. (2010) Young children as protagonists and the role of participatory, visual methods in engaging multiple perspectives, *American Journal of Community Psychology*, 46: 115–23.

Clark, A. and Moss, P. (2005) *Spaces to Play: More Listening to Young Children using the Mosaic Approach*. London: National Children's Bureau.

Clark, H. (2002) *Building Education: The Role of the Physical Environment in Enhancing Teaching and Research*. London: Institute of Education.

Commission for Architecture and the Built Environment (CABE) (2006) *Assessing Secondary School Design Quality*. Research report. London: CABE.

Cooper, I. (1981) The politics of education and architectural design: The instructive example of British primary education, *British Educational Research Journal*, 7(2): 125–36.

Cooper, I. (1985) Teachers' assessments of primary school buildings: The role of the physical environment in education, *British Educational Research Journal*, 11(3): 253–69.

den Besten, O., Horton, J. and Kraftl, P. (2008) Pupil involvement in school (re)design: Participation in policy and practice, *Co-Design*, 4(4): 197–210.

Department for Business, Innovation and Skills (BIS) (2013) *Construction 2025 Industrial Strategy: Government and Industry in Partnership*. HM Government.

DfES (2002) *Time for Standards: Reforming the School Workforce*. DfES/0751/2002. London: DfES.

DfES (2003a) A New Specialist System: Transforming Secondary Education, DfES/0173/2003. London: DfES.

DfES (2003b) *Building Schools for the Future: Consultation on a New Approach to Capital Investment*. London: DfES.

DfES (2003c) Classrooms of the future: Innovative designs for schools. www.education.gov.uk/publications/eOrderingDownload/DfES-0162-2003.pdf.

DfES (2005) *Harnessing Technology – Transforming Learning and Children's Services*. DfES 1296–2005DOC-EN. www.dfes.gov.uk/publications/e-strategy

Dudek, M. (2000) *The Architecture of Schools: The New Learning Environments*. London/New York: Routledge.

Dyson, A., Howes, A. and Roberts, B. (2004) What do we really know about inclusive schools? A systematic review of the research evidence, in D. Mitchell (ed.), *Special Educational Needs and Inclusive Education: Major Themes in Education*. London: RoutledgeFalmer.

Earthman, G.I. (2004) Prioritisation of 31 criteria for school building adequacy. American Civil Liberties Union Foundation of Maryland. www.aclumd.org/aTop per cent20Issues/Education per cent20Reform/EarthmanFinal10504.pdf

Eastman, C.M., Eastman, C., Teicholz, P., Sacks, R. and Liston, K. (2011) *BIM Handbook: A Guide to Building Information Modelling for Owners, Managers, Designers, Engineers and Contractors*. Canada: John Wiley and Sons.

Egan, J. (1998) *Rethinking Construction: Report of the Construction Task Force*. London: HMSO.

Engeström, Y. (1999) Innovative learning in work teams: Analysing cycles of knowledge creation in practice, in Y. Engeström et al. (eds), *Perspectives on Activity Theory*. Cambridge: Cambridge University Press, pp. 377–406.

Engeström, Y. and Ahonen, H. (2001) On the materiality of social capital: An activity-theoretical exploration, in H. Hasan, E. Gould, P. Larkin and L. Vrazalic (eds.), *Information Systems and Activity Theory. Vol. 2: Theory and Practice*. Wollongong, NSW, Australia: University of Wollongong Press, pp. 115–29.

Engestrom, Y. and Middleton, D. (ed.) (1996) *Cognition and Communication at Work*. Cambridge: Cambridge University Press.

Fisher, K. (2005) Research into identifying effective learning environments. Paper presented at the First OECD ad hoc Experts' Meeting on Evaluating Quality in Educational Facilities, Lisbon. www.oecd.org/education/country-studies/centreforeffectivelearningenvironmentscele/37905387.pdf

Foxell, S. and Cooper, I. (2015) Closing the policy gaps, *Building Research and Information*, 43(4): 399–406.

Galasiu, A.D. and Veitch, J.A. (2006) Occupant preferences and satisfaction with the luminous environment and control systems in day lit offices: A literature review, *Energy and Buildings*, 38: 728–42.

Gathorne-Hardy, F. (2001) Inclusive design in schools, *Support for Learning*, 16(2): 53–5.

Giddens A. (1984) *The Constitution of Society*. Cambridge, MA: Polity Press.

Gieryn, T. F. (2002) What buildings do, *Theory and Society*, 31: 35–74.

Gieryn, T.F. (2000) A space for place in sociology, *Annual Review of Sociology*, 26: 463–96.

Gislason, N. (2010) Architectural design and the learning environment: A framework for school design research, *Learning Environments Research*, 13: 127–45.

Gislason, N. (2015) The open-plan high school: Educational motivations and challenges, in P. Woolner (ed.), *School Design Together*. Abingdon: Routledge, pp. 101–20.

Goldacre, B. (2013) *Building Evidence into Education*. London: Department for Education.

Hargreaves, D. (2003) *Education Epidemic: Transforming Secondary Schools through Innovation Networks*. DEMOS.

Harty, C. and Whyte, J. (2010) Emerging hybrid practices in construction design work: Role of mixed media, *Journal of Construction Engineering and Management*, 136(4): 468–76.

Heppell, S., Chapman, C., Millwood, R., Constable, M. and Furness, J. (2004) *Building Learning Futures: A Research Project at Ultralab within the CABE/RIBA 'Building Futures' Programme*. London: CABE.

Higgins, S., Hall, E., Wall, K., Woolner, P. and McCaughey, C. (2005) *The Impact of School Environments: A Literature Review*. Design Council Report.

Hillier, B. and Hanson, J. (1984) *The Social Logic of Space*. Cambridge: Cambridge University Press.

Hodson, P., Baddeley, A., Laycock, S. and Williams, S. (2005) Helping secondary schools to be more inclusive of Year 7 pupils with SEN, *Educational Psychology in Practice: Theory, Research and Practice in Educational Psychology*, 21(1): 53–67.

House of Lords Select Committee on National Policy for the Built Environment (2016) *Building Better Places*. Report of Session 2015–16.

Hygge, S. (2003) Classroom experiments on the effects of different noise sources and sound levels on long-term recall and recognition in children, *Applied Cognitive Psychology*, 17: 895–914.

James, S. (2011) *Review of Education Capital*. London: Department for Education.

Johnson, P., Fisher, K., Gilding, T., Taylor, P.G. and Trevitt, A.C.F. (2000) Place and space in the design of new learning environments, *Higher Education Research and Development*, 19(2): 221–37.

Kerosuo, H. (2015) BIM-based collaboration across organisational and disciplinary boundaries through knotworking. *Procedia Economics and Finance*, 21: 201–8.

Kraftl, P. (2012) Utopian promise or burdensome responsibility? A critical analysis of the UK government's building schools for the future policy, *Antipode*, 44(3): 847–70.

Latham, M. (1994), *Constructing the Team*. London: HMSO.

Leadbeater, C. (2004) *A Summary of 'Personalisation through Participation': A New Script for Public Services*. DEMOS.

Leiringer, R. and Cardellino, P. (2011): Schools for the twenty-first century: School design and educational transformation, *British Educational Research Journal*, 37(6): 915–34.

Luckin, R. (2010) *Re-designing Learning Contexts: Technology-Rich, Learner Centred Ecologies*. London: Routledge.

Mahony, P. and Hextall, I. (2013) 'Building schools for the future': 'Transformation' for social justice or expensive blunder? *British Educational Research Journal*, 39(5): 853–71.

Mayrowetz, D. and Weinstein, C. (1999) Sources of leadership for inclusive education: Creating schools for all children, *Educational Administration Quarterly*, 35(3): 423–49.

Microsoft Consulting Services (2005) *Building Schools for the Future: An Opportunity to Personalise Learning and Fundamentally Re-think the Business of Education*. Report prepared for Kent County Council.

Moos, R.H. (1979) *Evaluating Educational Environments: Procedures, Measures, Findings and Policy Implications*. San Francisco, CA: Jossey-Bass.

Morgan, J. (2000) Critical pedagogy: The spaces that make the difference, *Pedagogy, Culture and Society*, 8(3): 273–89.

Morrell, P. (2015) *Collaboration for Change*. London: Edge Publications.

OECD (2013) *Innovative Learning Environments*. Paris: OECD.

OECD (2014) *Effectiveness, Efficiency and Sufficiency: An OECD Framework for a Physical Learning Environments Module*. Paris: OECD.

Parnell, R., Cave, V. and Torrington, J. (2008) School design: Opportunities through collaboration, *CoDesign: International Journal of CoCreation in Design and the Arts*, 4(4): 211–24.

Porter, J. (2016) Time for justice: Safeguarding the rights of disabled children, *Disability and Society*, 31(8): 997–1012.

PricewaterhouseCoopers LLP (2007) *Evaluation of Building Schools for the Future – 1st Annual Report*. London: DCSF.

PricewaterhouseCoopers LLP (2008) *Evaluation of Building Schools for the Future – 2nd Annual Report*. London: DCSF.

PricewaterhouseCoopers LLP (2010) *Evaluation of Building Schools for the Future – 3rd Annual Report*. London: DCSF.

Prosser, J. (2007) Visual methods and the visual culture of schools, *Visual Studies*, 22(1): 13–30.

Radcliffe, D., Wilson, H., Powell, D. and Tibbetts, B. (2008) *Designing Next Generation Places of Learning: Collaboration at the Pedagogy-Space-Technology Nexus*. Sydney: Australian Learning and Teaching Council.

RIBA (2017) *Better Spaces for Learning*. www.architecture.com/knowledge-and-resources/resources-landing-page/better-spaces-for-learning

Rutter, M. (1979) *Fifteen Thousand Hours: Secondary Schools and Their Effects on Children*. London: Open Books.

Saltmarsh, S., Chapman, A., Campbell, M. and Drew, C. (2015) Putting 'structure within the space': Spatially un/responsive pedagogic practices in open-plan learning environments, *Educational Review*, 67(3): 315–27.

Shaughnessy, R.J., Haverinen-Shaughnessy, U., Nevalainen, A. and Moschandreas, D. (2006) A preliminary study on the association between ventilation rates in classrooms and student performance, *Indoor Air*, 16(6): 465–8.

Sigurðardóttir, A.K. and Hjartarson, T. (2011) School buildings for the 21st century: Some features of new school buildings in Iceland, *Center for Educational Policy Studies Journal*, 1(2): 25–43.

Sutherland, R. and Fisher, F. (2014) Future learning spaces: Design, collaboration, knowledge, assessment, teachers, technology and the radical past, *Technology, Pedagogy and Education*, 23(1): 1–5.

The Key (2015) www.thekeysupport.com/about/media-press/school-buildings/

Tse, H.M., Learoyd-Smith, S., Stables, A. and Daniels, H. (2014) Continuity and conflict in school design: A case study from Building Schools for the Future, *Intelligent Buildings International*, 7(2–3), 64–82.

Tufvesson C. and Tufvesson J. (2009) The building process as a tool towards an all-inclusive school, *Journal of Housing and the Built Environment*, 24: 47–66.

Tuohy, P.G. and Murphy, G.B. (2015) Are current design processes and policies delivering comfortable low carbon buildings? *Architectural Science Review*, 58(1): 39–46.

Vasagar, J. (2012) Buried report praised Labour's school building programme, *Guardian online*. www.theguardian.com/education/2012/jul/05/buried-report-labour-school-building

Veloso, L., Marques, J.S. and Duarte, A. (2014) Changing education through learning spaces: Impacts of the Portuguese school buildings' renovation programme, *Cambridge Journal of Education*, 44(3): 401–23.

Victor, B. and Boynton, A. (1998) *Invented Here: Maximising Your Organisation's Internal Growth and Profitability*. Boston: Harvard Business School Press.

Volk, R., Stengel, J. and Schultmann, F. (2014): Building Information Models (BIM) for existing buildings – Literature review and future needs, *Automation in Construction*, 38: 109–27.

Vygotsky, L.S. (1987) *The Collected Works of L.S. Vygotsky. Vol. 1: Problems of General Psychology, Including the Volume Thinking and Speech*, ed. R.W. Rieber and A.S. Carton, trans. N. Minick. New York: Plenum Press.

Watts, J. (1977) *The Countesthorpe Experience*, London: Allen and Unwin.

Winterbottom, M. and Wilkins, A. (2009) Lighting and discomfort in the classroom, *Journal of Environmental Psychology*, 29(1): 63–75.

Woolner, P. and Thomas, U. (2016) Change and stasis within design and practice over three decades in an English primary school. ECER, Dublin, 23–6 August.

Woolner, P., Hall, E., Wall, K., Higgins, S., Blake, A. and McCaughey, C. (2005) School Building Programmes: Motivations, Consequences And Implications. Research report. Reading: CfBT.

Woolner, P., Hall, E., Higgins, S., McCaughey, C. and Wall, K. (2007) A sound foundation? What we know about the impact of environments on learning and the implications for Building Schools for the Future, *Oxford Review of Education*, 33(1): 47–70.

Woolner, P., McCarter, S., Wall, K. and Higgins, S. (2011) Changed learning through changed space. When can a participatory approach to the learning environment challenge preconceptions and alter practice? Paper presented at AERA.

Woolner, P., McCarter, S., Wall, K. and Higgins, S. (2012) Changed learning through changed space: When can a participatory approach to the learning environment challenge preconceptions and alter practice, *Improving Schools*, 15(1): 45–60.

Young, M. and Muller, J. (2010) Three educational scenarios for the future: Lessons from the sociology of knowledge, *European Journal of Education*, 45(1): 11–27.

Credit: HKS Architects

3
THE AIMS, SCOPE AND METHOD OF THE *DESIGN MATTERS?* PROJECT

Introduction

Limited research on the working practices of the construction industry suggests a better understanding of current practices is required before significant innovation and improvement can take place. The limited research in this area also fails to offer a systematic analysis of how space is shaped (Woolner et al., 2007) and there is very limited research which explores how space is shaped from the inception of a design idea through to occupation and the mediating effects of this space on end-users. One of the major challenges of the *Design Matters?* project was to develop a methodology for systematically analysing the relationship of school space to the experiences of students, teachers and parents. In this chapter we will discuss the methodology that we have developed in the course of this project by, first, giving background information on the previous POE literature, philosophical and theoretical underpinning of *Design Matters?*, second, outlining research aims and scope, and third, describing the data collection and analysis methods in detail.

Background

Design Matters? is a multi-disciplinary collaboration between academic educationalists and practising architects. The formulation of the project began with discussions about finding ways of understanding the impact of the school environment that draw on a much richer range of data than much of the existing literature. Daniels' particular interest is in schools as activity systems (e.g. Daniels, 2001) in addition to previous published research into the function and value of wall displays (Daniels, 1989). As an activity system, a school is a holistic entity, the various aspects of which work in mutual relations towards broadly agreed social purposes. Stables' interest is in schools as signifying environments, or semiotic systems, drawing on his earlier work on educational organisations as 'imagined communities in discursive space' (Stables, 2006, 2009[2003]). Both were therefore interested in how the visual and physical aspects of a school interact with the lived experience of students and teachers.

Both activity theory and semiotics reject the idea that the spatial environment should be studied in a way that is completely at odds with approaches to psychological and sociological aspects of school processes, such as teaching effectiveness or learner identities. Design must impact on these in some way (there cannot surely be no difference between studying in a bunker and in a forest, for example), but the degree to which, and the ways in which, design 'matters' remain under researched, notwithstanding the body of excellent work alluded to in the previous chapter.

Several months of exploratory discussions resulted in a successful application to the EPSRC (UK Engineering and Physical Science Research Council) for a Knowledge Transfer Champion grant to support Tse to collaborate further with us on developing a much broader proposal. Knowledge Transfer Champions were funded to bring together academics and practitioners in order to strengthen the links between academic research and professional practice. This period of reflection, networking, discussion and exploration was crucial to the development of the *Design Matters?* bid to the AHRC (UK Arts and Humanities Research Council) that was completed in late 2011 and approved for commencement in autumn 2012.

Rationale: a philosophical problem behind a research problem

There is a philosophical schism within Western Enlightenment thought that has strong, sometimes unconscious, effects on how we tend to research. Central to modernist mainstream thinking is a sort of Cartesian dualism that tends to separate the activities of human mind, operating in terms of reasons and freedom of the will, and biophysical nature, operating in terms of causes and fixed universal laws. Under the influence of this philosophical dualism, we tend to research human experience in terms of narrative, emotion, motivation and explanations, while we research 'space' as impersonal, law-driven and impervious to narrative, emotion, motivation and explanation. In the case of post-occupancy evaluations (POEs) of schools and other buildings used as dwellings (as discussed in Chapter 1), the emphasis on building performance tends to be restricted to important but limited aspects such as energy use, and air and light quality. At the same time, analyses of students', teachers' and schools' performance tend to focus solely on outcomes such as examination results and truancy rates. It is as though design and practice are regarded as different 'forms of life' (to use Wittgenstein's term: Wittgenstein, 1967) and are described by means of different discourses. Thus it is hard to form a rich picture of the complex interactions between them.

While *Design Matters?* is not of itself a philosophical project, we take on board this insight in arguing that it is no longer justifiable to evaluate the physical environment by means that are totally at odds with those used to evaluate human experience with its dimensions of affect, reason and narrativity; hence the importance of understanding design in relation to practice, as argued in Chapter 1. We are not thinking, feeling beings moving through an inert and mechanical physical space; rather, we are 'Here-Now'. (For a fuller discussion of this position, see Stables, 2012.)

Whatever the effects of school design on the experience of schooling, therefore, such effects cannot be reduced to the easily measurable aspects of environmental performance that architects and engineers have focused on in the post-occupancy evaluations of school buildings to date: aspects such as energy efficiency and air quality, for example. At the same time, educational experience cannot be divorced from such considerations. For example, Mumovic et al. (2009) have shown how levels of carbon dioxide build-up in classrooms affects mental acuity, while Barrett et al. (2015) have provided the most detailed insight to date into how features of classroom design can impact on the efficiency of teaching and learning, although that study was unable to take into account teacher performance. The field of post-occupancy evaluation has provided direction on how evidence can be gathered about the performance of educational facilities for over 50 years (Foxell and Cooper, 2015). Based on a critical review of the literature on the post-occupancy evaluation of physical learning environments, Cleveland and Fisher (2014) demonstrated that the majority of tools developed in various countries to evaluate school learning environments focus predominantly on the physical features of the physical environment itself; for example, Sanoff's multiple evaluation tools (Sanoff, 2001), Design Quality Indicators for Schools (DQIfS) (CABE, 2005), Educational

Facilities Effectiveness Instrument (EFEI) (FNI, 2011). Cleveland and Fisher (2014) formed the conclusion that POEs on the social or human components of the learning environment in supporting pedagogical activities are in their infancy and require further development.

In more recent years, the physical environment still seems to be the focus of school POE (e.g. Burman, 2016; Pamer et al., 2016; Burman et al., 2018). For example, Burman et al. (2018) did a building performance on five educational buildings constructed over the period 2007–2010 in England under the BSF programme and concluded that the schools are not achieving the energy performances of design intentions.

In more recent years, the body of POE knowledge has been expanded by, for example, an interdisciplinary theoretical framework developed by Baars et al. (2018) regarding the relationship between the psychosocial learning environment (PSLE) and the physical learning environment (PLE). However, further research is needed to test the preliminary PPR framework in an empirical setting to prove the validity, usability and reliability of the framework. Our challenge, therefore, was to find a way of researching school experience that gave increased insight into the mutual shaping of design and practice, given the complexity and dynamic nature of this interrelationship.

The Design Matters? challenge: from space to place (to home?)

In planning the project, we therefore began with two important assumptions. First, something spatial, such as a school building, is unlikely simply to 'cause' certain outcomes in terms of human behaviour, though it is equally unlikely not to affect such outcomes in some way. 'Shaping' does not necessarily imply causation. In an extreme situation, the physical environment may seem almost to determine human activity, as in falling off a cliff edge if you walk towards it without looking where you are going, or being caught up in an explosion. Even in these extreme cases, however, where one finds oneself, or what one is doing, or what level of protection one's clothing affords are important factors. Conversely, one can imagine situations in which one's environment seems to have very limited effect: in the immediate aftermath of a domestic crisis, for example, one is likely to feel unhappy and anxious wherever one is. In general, space and practice have some role in mutually shaping each other: what we do affects where we do it, and where we do it affects what we do. This relationship is complex and not directly deterministic. It is also under-researched.

Second, the role of design becomes increasingly difficult to determine as occupation develops. We can think of this in terms of Space becoming Place (the room designed to be a science laboratory becomes one in practice) and then Place becoming more like 'Home' (i.e. the science laboratory becomes a place in which students and teachers feel comfortable and empowered, one hopes). In other words, as people begin to utilise a building, they appropriate it for their own ends, and the response of the design to this appropriation is an important area to study.

It is actually very difficult to draw a firm line at the point at which 'design' ends and 'occupation' begins. Take the example of an airport terminal. Such a building is often 'designed' as a large, cavernous, almost empty internal space, yet it is also 'designed' to allow for strict surveillance and the tight management of person-flow. Thus our experiences of being in departure lounges are characterised more by crowded queues between plastic fences than by dwelling in large open space. The same often applies to schools, where students are often formed into lines and where access to certain areas is subject to strict restrictions and regulation. In any individual case, this regulation may or may not have been part of the design intention. As we shall see through many examples throughout this book, new-build schools were often designed specifically to allow both a greater sense of community and increased teacher surveillance. How these tensions play out in particular circumstances is not determined by, though it is affected by, the architecture.

Origins of Design Matters?

We were faced with challenge of developing an approach to the analysis and description of schools which allowed us to progress an analysis of the consequences of different types of design. Our argument was that in different schools (or cultures) actions and objects signify different meanings. We turned, inter alia, to previous work on wall displays as tacit relays of underlying pedagogic priorities (Daniels, 1989).

This was a study of cultural transmission at the level of the selection and arrangement of images at the classroom level. The wall display study took measures of school modality. A connection was made between the rules the children used to make sense of their pedagogic world and the modality of that world. We took this evidence of cultural transmission with reference to wall display as a spur to the way in which cultural transmission may, or may not, be evidenced at the level of school design.

At a very general level it is possible to conceive of cultures or schools as worlds of signs and signs about signs (Hawkes, 1977). (For a fuller discussion of schooling as semiotic engagement, see Chapter 6.) In a sense adapting to cultural change is a process of adapting to changing systems of signification. For a child, moving from home to school is itself an act of cultural change and, for some, entails culture shock. That which is taken to signify competence in one culture may signify incompetence in another or irrelevance in a third. How then does a school transmit to children the criteria that are taken to signify appropriate learning? What are the cues offered to children in their attempts to read the signs of schooling? In the early day study it was argued that art displays are part of the system of signs that constitute the culture of schools, that through these acts of publicity the principles which regulate the curriculum are realised.

In many schools to have a 'nice bright classroom with lots of good display work' is one of the commonly held indicators of good teaching practice. Display work is important not only to parents but also to children. Children like having their work displayed on the wall. This very public way in which a teacher shows approval of a child's activity is highly valued. By putting students' work on the wall the teacher is telling the child that he/she approves of it and at the same time is offering a model of good practice to the rest of the class. This, of course, is one of the reasons why children feel so proud when their work is displayed: their friends are being offered their work as a model. The way in which work is selected for display and indeed the way in which the display is arranged is effectively an act of publicity of the teacher's desired model of good practice.

Gearhart and Newman argued that, for the nursery school children they studied, learning the social organisation of a classroom and learning its curriculum could not be distinguished:

> What children know about drawing is intimately tied to what they . . . understand of drawing activities undertaken in a particular social . . . context.
>
> *(1980, p. 183)*

They discussed the importance of the way the teacher spoke to the children about their drawings and also drew attention to the particular form of pedagogy in the classroom:

> Drawing was also being learned from the teacher's efforts to teach the organisational independence of individual production tasks. Reflexively, this individual task organisation was being learned from the teacher's efforts to teach independently planful drawing.
>
> *(Gearhart and Newman, 1980, p. 183)*

The aims, scope and method of the *Design Matters?* project **37**

While Gearhart and Newman's study is of interest, it failed to undertake the comparative work needed to show how ways of learning to draw differ under different forms of classroom social organisation. Also, following as it does an explicitly Vygotskian experimental approach, it lacks the potential for describing and analysing the social organisation of the classroom in structural terms (Wertsch, 1985). In its failure to do this it confines interpretation to a very local domain. Through focusing on wall display rather than pupil – teacher and teacher – pupil verbal communications, a wider perspective on semiotic mediation was being drawn. If a wall display acts as a relay of tacit meanings, does school design act in the same way? How does the whole school environment communicate what counts as important in a particular school – if anything? On entering a school does the design itself help the student and/or the teacher to the answer the question 'What goes here?'

The photographs that are to be discussed here are representative of each school's display work. All the work displayed at one time in both schools was recorded and selected examples are presented. The selection was made by the teachers of the classes of 9–12-year-old children in each school. That is the (two) teachers in each school were shown the entire sample of photographs for their school and asked to select the three that best represented the school's display work. Emphasis was laid on the display rather than the individual pictures. Equally important is the fact that all the teachers responsible for this display work viewed their efforts as the result of a 'common sense' approach to the task. They did not regard themselves as having been instructed or coerced to work in this way nor did they regard their work as potentially different in form from display work in any other school. These photographs are those displayed in Figures 3.1, 3.2, 3.3, 3.4, 3.5 and 3.6.

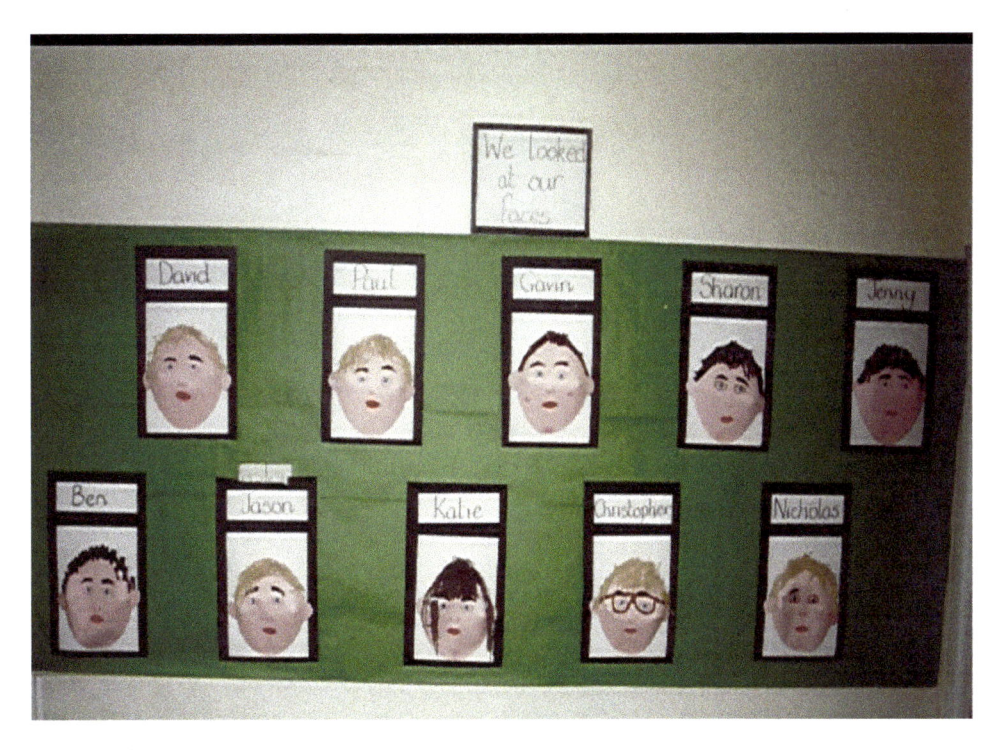

FIGURE 3.1 School display work

FIGURE 3.2 School display work

FIGURE 3.3 School display work

FIGURE 3.4 School display work

FIGURE 3.5 School display work

FIGURE 3.6 School display work

Interpretation of displays

What then is revealed by an inspection of a sample of the display work in these schools? The control over what is expected is clearly high in displays 3.1, 3.2 and 3.3. In 3.1 the faces all have the same structure – they are all the same shape! In 3.2 the faces of the flowers are structurally similar. The faces were all yellow, all on the same plates, all with red lips and all had eyebrows. The levels of similarity in 3.3 are so marked that they require no comment.

On the other hand, the control over what is taught/expected is of a very different nature in 3.4, 3.5 and 3.6. In 3.4 there is an integrating theme of transport and yet children have produced different illustrations relating to the central theme. These are drawn, crayoned or painted using a variety of techniques. In 3.5 and 3.6 there are no underlying themes and the work is very varied in terms of the techniques used and the content portrayed. It seems there are at least two principles at this level of control which distinguish the schools. In one school there is a high degree of control over what is to be portrayed and also over the techniques and materials to be used. In the other school, the level of control over these factors is much lower.

It is perhaps worth considering the relation of the conceptual foci of two of these displays. The concept underlying display 3.3 is that of letter recognition and this is explicitly noted in the labelling. The implicit concept underlying 3.4 is of a different order – transport. It may be that this reveals different theories of curriculum sequencing. On the one hand, a holistic strategy is revealed in the integrated approach of the theme transport and, on the other hand, an atomistic strategy, that of a phonic approach to the teaching of reading, is implied. This is reminiscent of a familiar debate. Displays 3.1, 3.2 and 3.3 appear to be in accord with the strategic principle advocated by Gagne (1985), who argued that children cannot understand complex ideas before they have mastered the notions which are more conceptually primitive. Whereas displays 3.4, 3.5 and 3.6 appear to reveal the strategy accorded to Bruner (1986), who argued that children will not understand and remember 'simple' ideas until they recognise the framework into which they fit.

Each school appears to some extent to have a characteristic style of structuring the displays. Whereas in 3.1 and 3.2 the pictures are arranged in straight lines with regular spacings between pictures, in 3.4, 3.5 and 3.6 the pictures are closely grouped in irregular patterns. It is perhaps not entirely coincidental that in picture 3.4 the work displayed was produced by children in the age range 5–14 where each display in the other school was produced by one age group only. These two factors perhaps reveal underlying levels of classification. On the one hand, ages and individuals are grouped and, on the other,,,, separated by clearly marked boundaries. This is illustrative of some of the ways in which it is possible to argue that the principles on which the curriculum is organised are realised in the way work is displayed.

The arrangements through the production, selection and combination of children's painting may be seen to act as a relay, or at least as potentially indicative of, the deep structure of the pedagogic practice of particular schools, although as far as the teachers were concerned, they were simply mounting wall displays rather than using wall displays explicitly as relays of the focus of their practice. While they were keen to create a good impression through their wall display work, they were not necessarily aware of their expression of the underlying principles of school practice.

Following the directions given by Vygotskian psychology, it was deemed profitable to investigate the meaning of wall displays for children as a step in the process of understanding what counts as important in a particular school (Wertsch, 1985). In the investigation of wall display, it is important to remember that the children also produced the pictures and thus were socialised by that activity. The products of these socialising activities are then selected, combined and organised by the teacher in a way which celebrates and announces the expected competences required of a particular school and/or classroom. Rather than reading backwards from statistics describing the outputs of schooling, it would seem worthwhile to consider what is relayed to children by particular activities.

From this perspective schools may be considered as generators of a specialised within-school semiotic code. The meaning of these signs for the participants in the practice of schooling then becomes the object of study. The study of wall displays indicated that children from different schools 'saw' different meanings in the same displays. They were oriented towards different sets of recognition and realisation rules. Here lies the possibility for the development of an interesting theoretical development which articulated the process of the internalisation of social/cultural factors mediated by sign systems is both theoretically exciting and of direct practical utility.

We shifted the focus of attention from wall display to the design of the school. We were concerned with the question as to the extent to which design transmits these messages or is it how that building is used in particular forms of practice. We were studying 11 year olds as they moved into newly built schools and we gathered data on the perceptions of their experiences of specific forms of pedagogic practice in specifically designed places.

The school as signifying environment: expectations, context and challenge

The impact of design on lived experience can be evaluated through various forms of post-occupancy evaluation each of which has its strengths and limitations. It can be argued that the practices of evaluating design in terms of a building's performance on objective measures is deeply embedded in the traditions of design and construction. However, variations in how buildings perform on standard measures are influenced by the ways in which the building is used. As we have already shown, one design may be used and perceived in very different ways depending on the occupying educational organisations, their values and beliefs and their practices. Individuals – affected, of course, by their cultural backgrounds and social positions – have differing expectations from schools and their responses are coloured by these.

This section therefore draws from the *Design Matters?* data to illuminate the variation in the ways that individuals can respond to a new school environment. To do so, it draws on some basic ideas from semiotics: specifically, from biosemiotics (Hoffmeyer et al., 2017) and edusemiotics (Stables and Semetsky, 2015). From a semiotic perspective, any environment is a set of significations, not merely a collection of objective entities arranged in space. For example, an ant might see a blade of grass as a path, while a cow would simply regard it as food.

School as signifying environment, or *umwelt*

Biosemiotics is specifically concerned with the interactions of organisms and their environments, and is founded in the work of the biologist Jakob von Uexkhuell, who referred to the signifying environment as the *umwelt*. Botar (2001) argues that von Uexkhuell was an inspirational figure for some early and mid-twentieth-century architects. Regardless of the extent of this influence, a school can be understood as an *umwelt* with its various spaces, places, practices and artefacts interpreted somewhat differently by different 'interpretive communities' (Fish, 1980) and the individuals within them. Thus we might expect, for example, that girls or boys, or children with different primary school experiences, or more or less academically successful students, might respond somewhat differently to any particular school environment. We might also expect that certain commonalities can be found within respondent groups, but that a degree of difference within groups will also emerge. The specific issue of trajectories from school to school were discussed in Chapter 5.

According to von Uexkhuell, an individual's mapping of his or her environment is that individual's *innenwelt*. In effect, the *innenwelt* amounts to a set of predispositions that orientate the individual in context, thus the concept bears some relation to Bourdieu's more explicitly sociological conception of an individual's *habitus*, the set of dispositions that enables someone to engage with a *field* (such as that of schooling in the broader sense). Another relevant concept here is the less explicitly semiotic one of *lebenswelt*: the human cultural *milieu*. As a school is the product of specifically human cultural choices, one can understand the *umwelt* of a particular school as (at least in part) nested within a particular *lebenswelt* or set of cultural expectations. (Thus, for example, we might expect to find different school designs in highly authoritarian administrative regimes, very liberal administrative regimes, or in societies with very strict codes of moral behaviour in areas such as gender roles.) Each new student enters the *umwelt* of a secondary school with an *innenwelt* or *habitus* moulded from previous experience, albeit usually, though not always, from the same general *lebenswelt* as those who created, and who manage, the school.

It would be tempting at this point to introduce a graphic showing how *innenwelt* is nested within *umwelt* within, in this case, *lebenswelt*. However, this would be misleading, as there is not one *umwelt*: the same school may not, in effect, be the same environment for each actor within it.

Semiotic codes and interpretations as readings

On this account, an environment such as a school is a sort of text that is interpreted by each individual and set of individuals in the light of previous experience (Stables, 2009[2003]). As there are different ways in which one can read and understand a written text, so there are different ways in which one can read and understand one's school. Note that the word 'text' derives from the Latin *textere*: to weave. An *umwelt* is a sort of interweaving of specific signs with their evident denotations and connotations.

Any semiotic system, or code, has two primary functions: to model the world (to create the *innenwelt*) and to communicate. The first is personal, the second interpersonal, though ultimately each relies on the resources of the other. As Wittgenstein argued, there can be no private language even though someone

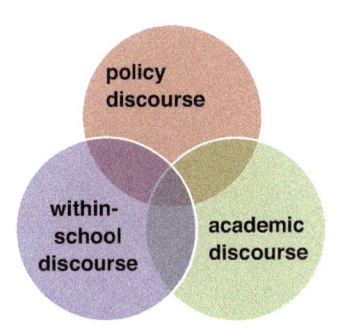

FIGURE 3.7 Overlapping semiotic codes of schooling

might live a largely isolated existence (Wittgenstein, 1967). There are two sides to communication of any sort: there are senders and receivers of messages.

In the case of schools, it can be argued that there are three overlapping types of semiotic code at work (Figure 3.7). The policy discourse of schooling comprises messages largely sent from politicians and interpreted by the general public, including teachers, students and parents. Within-school discourse comprises largely messages sent by teachers and interpreted by students, and is thus likely to have the greatest direct effect on students. Recent research into how teachers and students operationalise key concepts such as 'effort' and 'ability' illustrates how each of these domains has distinctive characteristics: they overlap but do not merely coincide, and within-school discourse is both more and less than a watered-down version of academic or policy discourse, while the latter two are also somewhat distinct (Stables, Murakami, McIntosh and Martin, 2014; Stables et al., 2018).

Considering the Sender role, we can see that a particular policy intention will be variously interpreted by school personnel. When the student perspective is considered, another level of variation comes into play. Each individual will encounter a situation as positive, negative or neutral, and will work with these messages relating to her social identity (as academic failure or success, for example, or as socially popular or isolated) in order to develop her personal identity. (See Tarasti's 'existential semiotics' (Tarasti, 2000), and also Rom Harré's work on personal and social identity and positioning (Harré 1983, 1998, 2012).) In mainstream literacy studies, one common way of analysing reader response is in terms of the sorts of literacy being developed; for example, functional, cultural or critical literacy (e.g. Williams and Snipper, 1990). One of the *Design Matters?* team has worked on extending these concepts into the broader field of environmental literacy (e.g. Stables, 1998; Stables and Bishop, 2001). On this account, functional literacy relates to one's capacity to name and know how to deal with aspects of one's environment at an instrumental level; for example, to see a 'Stop' sign and to be able to stop. Cultural literacy relates to the capacity to understand the social significance of something, as in Hirsch's '100 things every American needs to know' (Hirsch, 1987). Critical literacy relates to the ability to account for current circumstances and argue how they might be changed: some argue that only critically literate students could be expected to engender positive social change.

Drawing these strands together, we can understand a particular school environment as one in which the teaching body, led by the management, sends a series of messages influenced, but not determined, by policy and research. The students interpret these messages on various levels and according to their prior experiences and habitual responses. The term 'messages' here does not, of course, refer only to the specific content of something spoken or written. Messages can be in any medium: signs, gestures, rules, even shapes, light levels and paint colours can all relay messages to a receiver – this is how road signs work, for example. Often, design plays an important role in messaging: the words delivered by a policymaker in a local authority do not have the same effect as the same words spoken by a homeless person on the street outside. The power of a message derives not only in what is said, but also in how it is said, where it is delivered, who delivers it, and what happened prior to its delivery. In different contexts, a manager's instruction to an employee to 'clear your desk' can mean different things.

Methodological considerations and implications

At the outset we were cautious of accounts of design as intervention that would change behaviour in the absence of patterns of social interaction and cultural influence that would take advantage of the possibilities that designs afford. As the headteacher of a newly built school remarked 'the design is a provocation to learn differently but it's what you do inside it that matters'.

While, in the design and build of a school, it is easy to envisage the commissioner of a school as the client, one must also consider the position of the student. The mutual shaping of person and place is recognised by Burke, who suggests that the 'vision of school as a transformed space for learning . . . could not exist separately from a transformation in the view of the child as artist of their own learning and builder of their own worlds' (Burke, 2010, p. 79). However, there are very few examples of a sociocultural analysis of school architecture as a structuring resource. Accordingly we set out to investigate the ways in which the design of spaces within schools mediates and shapes practices of teaching and learning.

In summary, we can expect considerable variation, but also considerable overlaps, between how individual children 'read' their school environments. Neither the design of a school nor any other feature of it, including its management culture, can totally determine how a student (or for that matter, a teacher or parent) responds to the school. These factors are, of course, important, however. For example, the design of a classroom offers certain invitations and certain constraints, to which people will tend to respond according to their preconceptions. Imagine, in this spirit, how a white, empty room with a single light and a couple of chairs might impact on someone approaching it for the first time: there will be a measure of trepidation, perhaps, or curiosity; these will vary in intensity from person to person. Against this, there are certain very strong messages that appeal almost universally: imagine the room has 'rest area' written over the door, or 'investigation room'. On the other hand, the sign 'prayer room' would elicit very varied responses. In short, the *umwelt* of the school is not quite the same for each occupier. This implies that, to a great extent, children entering secondary school will evaluate their initial experiences with respect to their last experiences of primary school (this particular issue will be addressed in the work on student trajectories in Chapter 5).

We sought to capture a sense of both the regularities and the idiosyncrasies of students' responses to the project schools through all the forms of qualitative data collection, including the nominal group techniques (NGTs). However, the NGT process, while revealing considerable variety, was intended to produce consensus within groups. In this chapter, therefore, we shall focus on the most open-ended of all the forms of data collection: the essay task.

Whereas the students always knew when doing the NGTs that they were contributing to a project about school design, the essay task was covert, conducted merely as a normal piece of business in an English lesson. The pupils were simply asked to create a piece of writing with the title 'My First Week at This School'. The resulting essays were marked and discussed by teachers as regular classroom business, with teachers drawing from the work whatever they wished. For the project team, however, this task had two particular advantages. First, the work was done individually, thus reducing any sense that dominant personalities might skew others' responses, or that shy students would not contribute. Second, there was absolutely no encouragement given to students to focus on design-related aspects of school life, so the emphasis given to such aspects should broadly reflect the importance of design in their initial experiences of a new school.

To this end, all the Year 7 (first year) students at 18 schools were asked to write a short essay entitled 'My First Week at This School', at or shortly after the end of the first week of the autumn term 2013. These comprised the 11 main project schools and seven of the comparator schools. As might be expected, the essays varied a great deal in terms of length and general levels of literacy, including breadth of vocabulary and accuracy of spelling, punctuation and grammar. However, the team's concern was simply with substantive content. To that end, essay data were independently coded by two researchers for emergent themes. This inductive analysis not only enabled the identification of themes but also made the data quantifiable and comparable across sources, as each thematic reference could be counted. Quantifying the emergent themes permitted the comparison of data between and across pupils, genders and schools. This method of co-ordinated data analysis reflects what is popularly regarded to be a systematic but flexible research strategy led by emergent categorisations (Punch, 2009). The findings are discussed in chapter 6.

Principles into practice: the overall project design

The project was therefore designed so that the interrelationships of design and practice could be explored as richly as possible. On this account, the effectiveness of design *is* the experience of design in practice, where 'practice' is understood in the broadest sense. The physical environment is not inert, nor, as a signifying environment (or *umwelt* in the semiotic tradition, after von Uexküll, 2010) is it even the same for all actors. (For example, a blade of grass can be a snack for a cow or a pathway for an ant.) It is misleading to think of the physical environment, even in its most unexploited form, as simply the entire collection of physical entities that surround us. Whether we are in a forest, a park, a school or an airport lounge, what we experience as our environment is conditioned by our evolution as a species and, to a lesser extent, by our cultural, and even individual histories. (Even sight is a product of evolution, and vision varies somewhat between species, as well as between human individuals.) Before a space designed as a classroom becomes a classroom, a visitor to that space may experience it differently from another visitor: as an extreme example, consider the likely contrasts between the reactions of a claustrophobic and an agoraphobic visitor to this space. As the space becomes a designated place and the site of practice, so such differences may multiply. It is impossible to shed one's cultural legacy in responding to a place of any sort. Places are what places seem, albeit one may be ignorant of how they 'seem' to others. Our challenge therefore was to understand the effects of design and practice on experience and vice versa as fully as possible. We therefore sought to balance our data collection in three important respects.

First, we sought to triangulate the perspectives of multiple actors: students, parents teachers, non-teaching staff, headteachers, commissioners, architects, engineers and contractors, Second, we sought to balance the largely qualitative data (for experience of living in a space is largely a matter of qualitative judgement) with some quantitative data. Third, we sought to juxtapose first-person (I/we) perspectives with third-person perspectives drawn largely from observations and secondary data analysis.

Aims and research questions

The aim of the project: to understand the impacts of the design of such schools on students', teachers' and parents' engagement in the educational process on a number of levels.

There were six specific research questions:

- RQ1. How did the design come to fruition? (Who commissioned it, and who was consulted, and about what? What informed the architects' proposals?)
- RQ2. What are children's expectations of the educational experience the new school will afford, prior to moving to the school and on entry to it?

- RQ3. How and why is the school used in the way that it is? (Are expectations realised in practice? How do students and teachers utilise the spaces and how do parents respond to the school? What would students and teachers design differently?)
- RQ4. What priorities are established through the above practices? (Why do such priorities emerge and how do they impact on learning and teaching behaviours and outcomes?)
- RQ5. How are the aims and objectives established in the design process reflected in the actions and perceptions of students and teachers?
- RQ6. What aspects of design would be desirable standard features?

Methodology

We employed a five-step, mixed methods design (Greene, 2007), collecting data through interview, observation and documentary analysis, thus capturing both first-person (subjective) and third-person (objective/intersubjective) perspectives over the phases of the project. We used Bernstein's work to develop an approach to the analysis and description of the schools as modalities of institutional practice which was used in the subsequent analysis of data concerning experiences of occupation as the designs were transformed under different theories of instruction as promoted by successive headteachers. In so doing it examines the relationships between the structuring of space in a building, the structuring of social relations and practices and the psychological consequences for occupants of the building. The following parts of this chapter will discuss (1) the sample, (2) the phases, (3) details of research instruments and (4) data analysis methods of the *Design Matters?* project.

The sample

The sample of schools consisted of 18 examples of different secondary designs including seven examples of 'traditional' schools used as comparators in our data analysis. The 11 new school designs in five English localities were commissioned by five local education authorities funded by the BSF and Academies Programme. The new schools were designed by four UK architectural practices and opened for occupation between 2003 and 2012, allowing the study to understand the implications of designs and educational practices after different periods of occupation.

As can be seen in Figure 3.8, perceptions data were gathered at primary school and at each occupation (new headteacher) of the secondary school. Alongside these perceptions data we observed the use of space in each school environment. These observations in turn informed the questions that were used in individual and group interviews. For example, if we observed differentials in the usage in the communal 'heart spaces' of the school we spoke with students and teachers about their perceptions of these spaces and their explanations for the ways in which they were used. We progressively refined our understanding of use and meaning of spaces through cycles of observation and interview. Not only did we progressively focus on emergent issues but we also followed students from their primary setting through any subsequent transformations of their secondary setting as changes in leadership were invoked by successive headteachers.

All the schools were located in areas of relative social deprivation, with higher than national average proportions of students entitled to free school meals (this is a common measure for school affluence and deprivation in the British context). The main sample of children comprised one class of Year 6 students from each of the secondary schools' feeder primaries: in total, $n = 293$. Children undertaking transfer from primary to secondary school. Additional students from the secondary schools partcipated in latter phases of the research (see Phase 3 below). Three of the new-build sample secondary schools have on-campus

Design Matters?

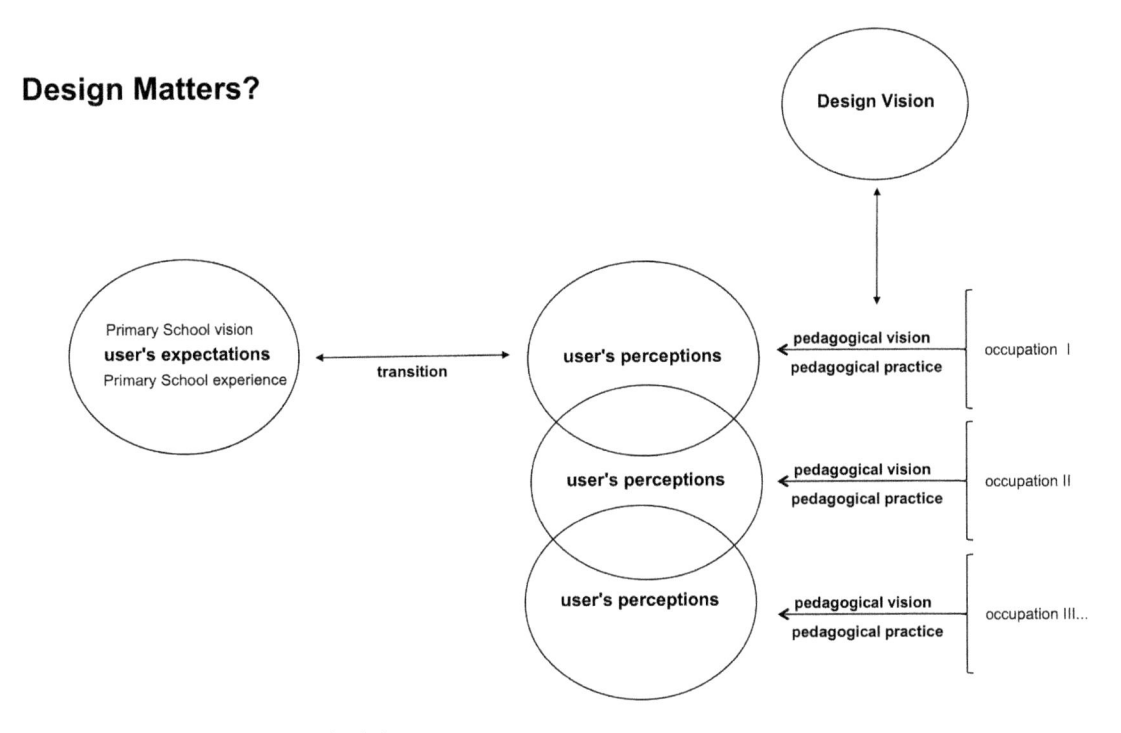

FIGURE 3.8 *Design Matters?* methodology

primary schools. We thus sampled transitions from new-build and comparator primary schools into new-build and comparator secondary schools. The schools have been anonymised, and are characterised throughout this text as follows:

- *Schools (area name) Locality A, B, C, D, E*: these comprise the main sample of new-build secondary schools A1, A2, A3, A4, etc. A descriptor for each appears below.
- *Comparator Schools (area) Locality A, B, C, D, E*: these comparator schools AC1, AC2, etc. were existing secondary schools from the same locality from which we collected comparative data.
- *Primary Schools (area) Locality*: these were the principal primary feeder schools AP1, AP2, AP3, AP4, etc. for the main sample schools.

The main sample of new-build secondary schools

School A1 is an Academy in a large town in the south of England; B1 is an Academy with Church of England sponsorship in a large town in the English Midlands; C1 is a comprehensive school (though in an area with academic selection to grammar schools at age 11) in a middle-sized southern English town; D1 is a former comprehensive school now with Academy status on the outskirts of a large town in south-west England, and E1 was one of the first academies built under the BSF initiative, in a west country city. All except E1 have undergone changes of headteachers since the project began, and B1, C1 and D1 have changed heads and management team two to three times during that period. This degree of instability should be borne in mind in all considerations arising from the *Design Matters?* project. What a school 'is' one year may be very different from the way it is a year or two later. Perhaps particularly

48 The aims, scope and method of the *Design Matters?* project

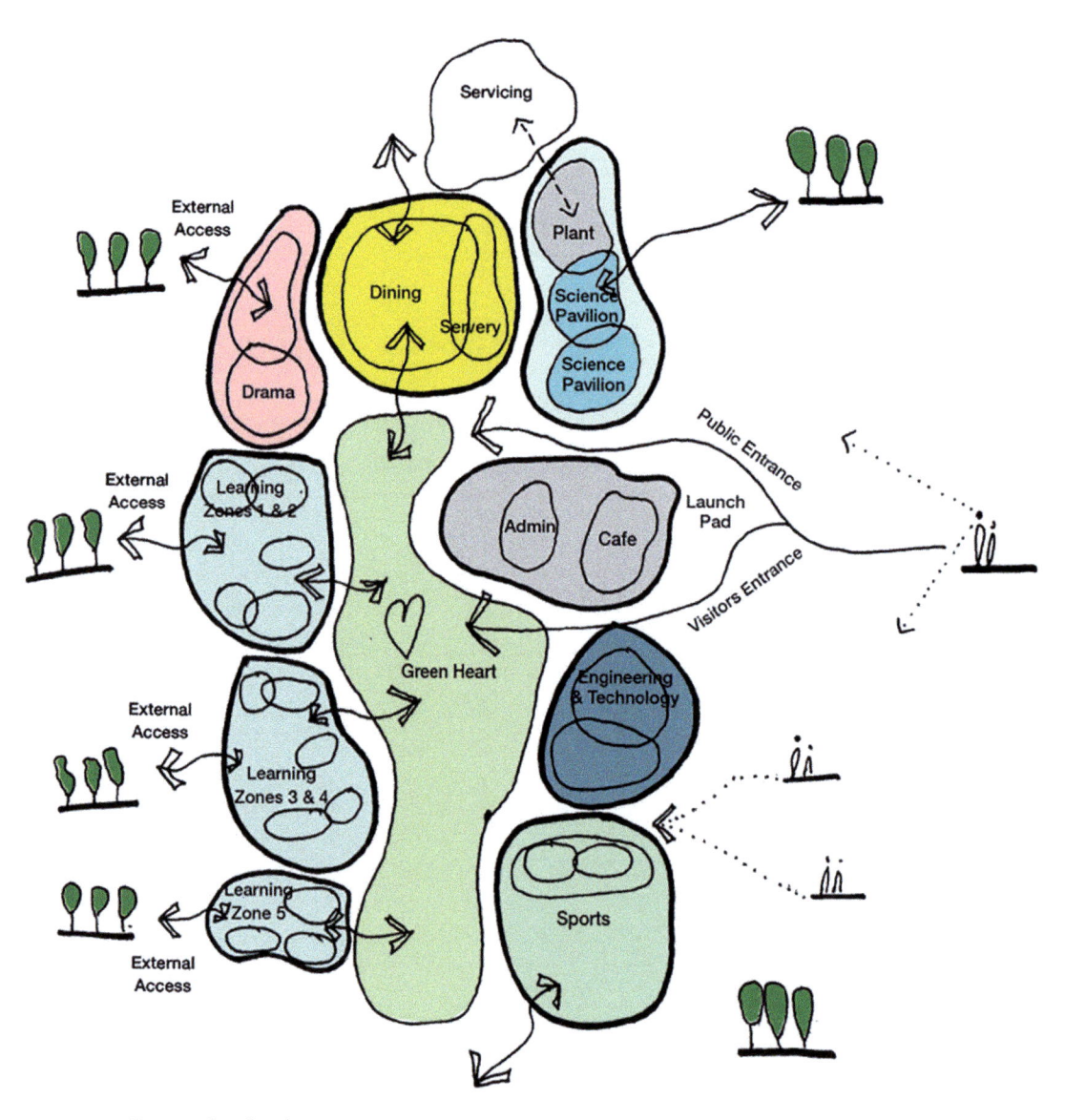

FIGURE 3.9 Case study school

in areas of challenging circumstances, schools can be radically reformed following negative inspection reports or even the retirement of previously successful headteachers. This high degree of contingency, contextuality and, often, associated instability is clearly a major issue for planners and policymakers that will be returned to in later chapters.

A1 is a smart, imposing building near the crest of a hill. It is characterised internally by very good site lines. The sense that this is a very 'adult' environment is reflected in the organisation of daily life within the school, which is characterised by tight control and rigid rules (e.g. uniform, movement) which have been reported to improve student behaviour and attainment, yet which have also caused a degree of resentment among some students.

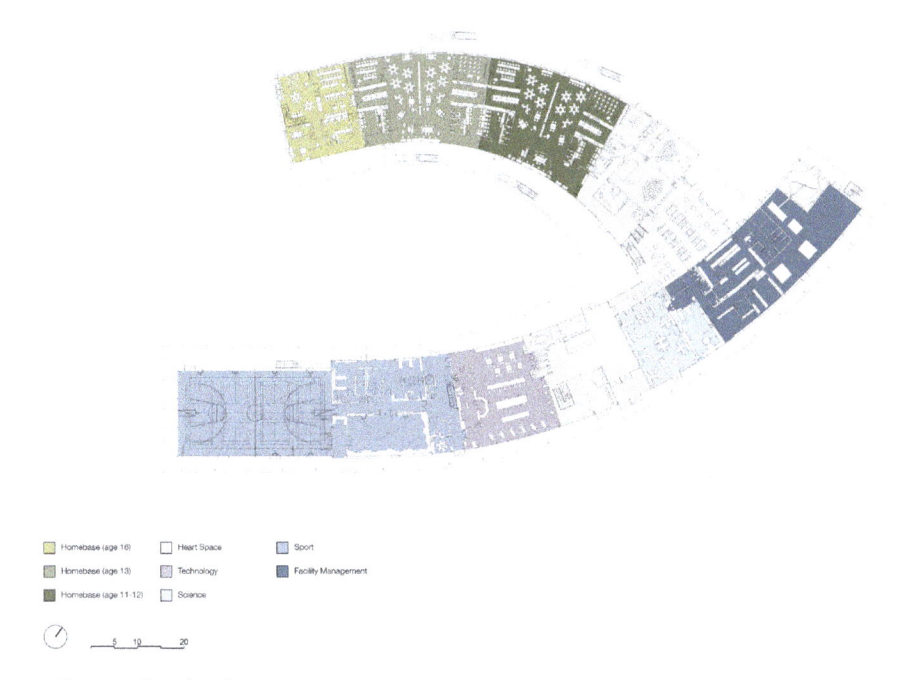

FIGURE 3.10 Case study school

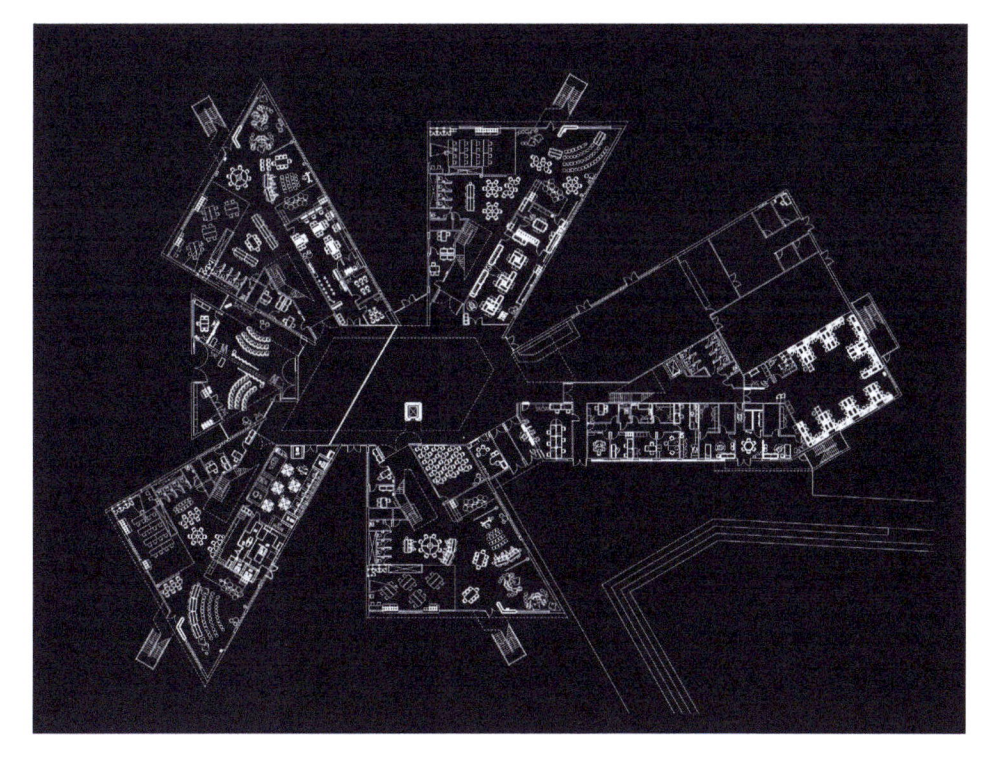

FIGURE 3.11 Case study school

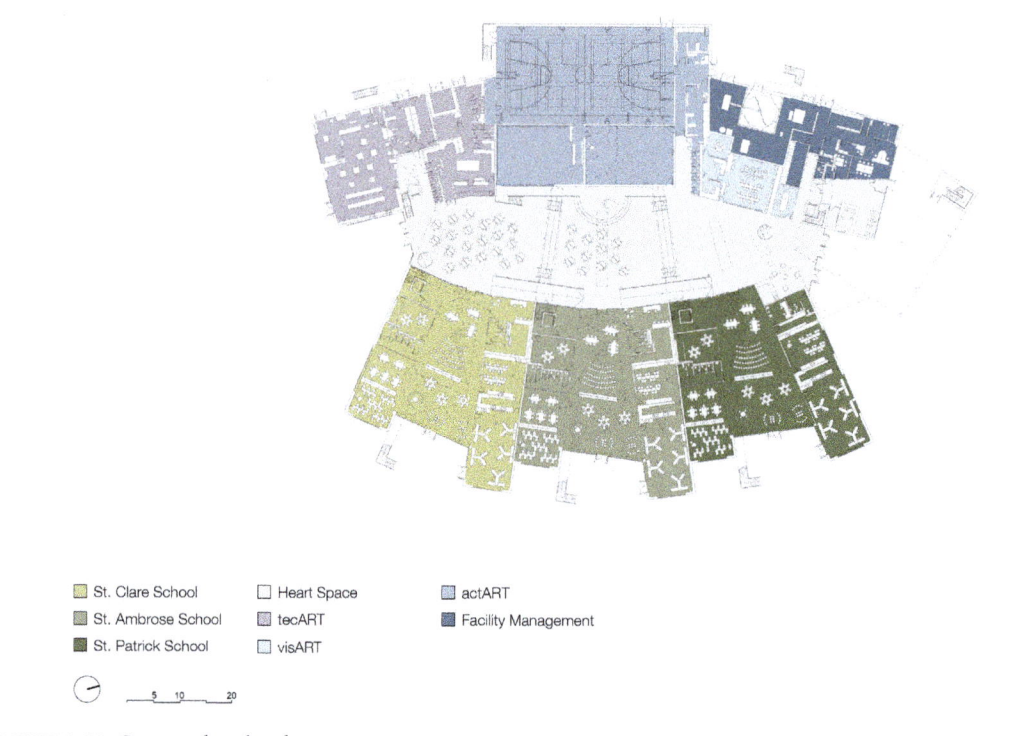

▨ St. Clare School	☐ Heart Space	▨ actART
▨ St. Ambrose School	▨ tecART	▨ Facility Management
▨ St. Patrick School	☐ visART	

FIGURE 3.12 Case study school

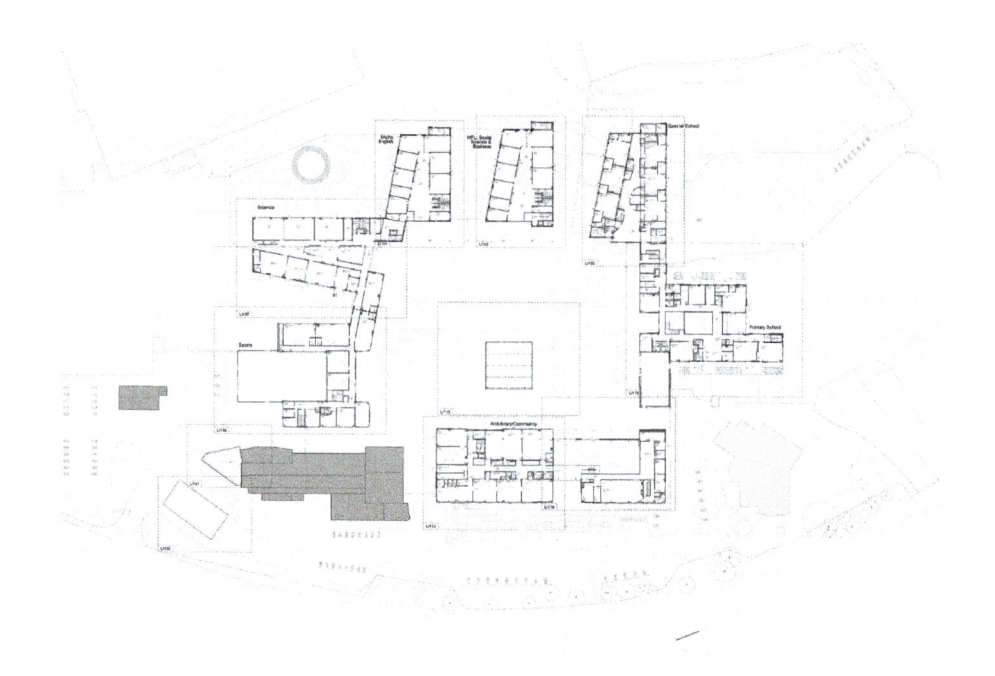

FIGURE 3.13 Case study school

B1 is characterised by an emphasis on pastoral care by the headteacher (2013–16), within a tightly controlled environment. The academy sponsor and the headteacher (2013–16) had a strong commitment to community transformation and with the design team created a building that invited the neighbourhood into the heart of the school. This school has had some success in improving examination results initially after occupation but had been identified as a cause for concern by the school inspectorate (Ofsted) during in 2015. This resulted in the resignation of the headteacher and changes of leadership and educational practices followed.

C1 has a large, open entrance area, or 'heart space' which also serves as a canteen/restaurant and is generally characterised by good sight lines and thus high visibility of students. At C1, many teaching spaces were designed on a large open-plan basis, with the intention that these spaces might be used flexibly and broken up as appropriate. Many of the larger spaces are not used to capacity because teachers feel the noise levels are too high if more than one class occupies the space. Despite (or perhaps because of) this lack of innovation, or rejection of the school's radical design possibilities, C1 features in the data as the first or second of the five schools in terms of student perceptions, and is enjoying steady progress in terms of examination success. Interestingly, although teachers have largely rejected the open-plan teaching spaces, C1 makes more use of the 'heart space' for individual and group work than is the case in the other schools.

D1 is the largest campus and feels generously spacious. There is widespread use of soft pastel colours and much evidence of student artwork in, for example, the public reception area. Also, there is the greatest evidence among the sample schools of continuity of personnel and, consequently, of educational vision during the long commissioning, design and build process.

E1, the oldest of the five schools, has undergone not only changes of leadership but also, along with many inner-city schools in the UK, a changing student demographic. It is a light, airy building, with 'communities' extending adjacent to a long 'street'-style main corridor. Originally designed as very open to the community, it has, like the other schools, become increasingly securitised, partly as a result of the greatly strengthened child protection agenda in the UK. The environment, while light and well resourced, is heavily controlled. For example, there is little scope for displays of student work or informal notices; instead, a series of rather corporate-seeming display boards prominently display positive messages about the school's mission and achievements.

The phases of the project

Phase 1 ran from November 2012 to June 2013. It comprised a literature review, an investigation of commissioning and design processes, and instrument development. Procedures for obtaining consent were negotiated with the schools. Research instruments were developed and trialled. This early engagement with the schools provided an opportunity to present an overview of the project and clarify any questions and concerns that the schools had with regard to the forthcoming project events.

During this phase, documentary evidence was sought from the design teams to inform our understanding of the design brief, the commissioning, design and construction processes (RQ1).

The design team involved in the design process, design commissioners and headteachers were interviewed in order to establish how the Pulham design came to fruition (RQ1). This involved consideration of the consultation process and the underlying ideas, theories and motives which were invoked by all parties in the design process. We also sought to understand any conflicts and trade-offs that emerged during this process.

In *Phase 2* (June to August 2013), we collected the data to answer RQ2 concerning children's expectations of the educational experience their new secondary school would afford. Interview groups were selected to comprise a representative range of the school population in terms of gender and prior academic achievement. A nominal group technique (NGT) was used; this, along with the other research tools, will be described in more detail in the following section. Brifely, NGT is a structured method for group brainstorming and establishing consensus that encourages contributions from all group members and fosters tolerance of conflicting ideas. It is particularly useful in situations when some group members are much more vocal than others or think better in silence.

In *Phase 3* (September 2013 to August 2014) we collected data to enable us to answer that part of RQ2 which refers to children's experiences of their new schools upon entry. We also undertook interviews, observations and documentary analysis in order to ascertain 'how and why is the school used in the way that it is' (RQ3/4/5). Reference to Phase 2 data and Phase 3 data together allowed us to consider how far, and in what way, expectations held by teachers, students and parents were being realised in practice. This also involved a consideration of how students and teachers utilised the spaces and how parents responded to the schools. Group discussions and interviews established what students and teachers would have designed differently. These also enabled us to identify the priorities established through the practices of the schools and to consider why they emerged and how they impacted on learning and teaching behaviours and outcomes.

In all but very few cases, the same students interviewed in Phase 2 were interviewed twice more (once before Christmas and once in the period between Easter and their summer holiday) in Year 7 (the first year at secondary school), thus providing both cross-sectional and longitudinal data (RQ3/4/5). We conducted semi-structured focus group interviews with teachers representative of their schools in terms of gender, subject expertise and years of service. Separate interviews were held with the headteacher of each school and senior staff responsible for teaching and learning and pastoral care (RQ1/3/4/5). A randomly selected group of parents was invited to take part in a short, structured interview, conducted as deemed appropriate in the context of each school (RQ4/5). One of these took place at a parents evening; the rest during the course of research team visits to the school.

Observations were made of classes in Years 7, 9 and 11, using observation schedules developed in Phase 1. These were further nuanced by both emergent theoretical expectations and initial data collection. Informal observations and photographs also informed our understandings of the schools' deployment of their space, and of how students respond to this (RQ3/4/5).

School facilities managers were interviewed with respect to their perceptions of design strengths and limitations. Documentary and statistical evidence were used to indicate trends in educational achievement (RQ4/5) and to inform other aspects of school use (relating to student behaviour, or provision for students with special educational needs, for example). Finally, Year 7 students were also asked to write open-ended accounts of 'my first week at . . . School'. These accounts were analysed by researchers to ascertain the role of design in such accounts and responses (RQ2), bearing in mind that students were not made aware that this exercise was part of the project or that the focus of readers' interest would be on design.

Phases 4 and 5 (September 2014 to June 2016) largely comprised data analysis (RQ3–6 in particular), re-engagement with the designers, feedback to the schools and early work on dissemination and impact. Feedback seminars were given at each of the five main project schools, with some attendees from other schools, those involved in the commissioning process and the architects.

Details of research instruments

Instruments used with students: first person

NGT (nominal group technique)

Students completed the NGT tasks in groups of about six. The procedure is:

1. Individuals generate ideas during or before the group meeting.
2. Each person takes a turn reading one of their ideas and ideas are written in a central place until all are listed.
3. The group discusses the ideas, possibly adding ideas to the list.
4. Each group member ranks the listed ideas.
5. Individual rankings are summarised for each idea to form a group ranking.
6. The group ranking of ideas is discussed (following Chapple and Murphy, 1996).

This method is designed to elicit data which are not overly determined by the structure or content of researcher questions and has a number of advantages. The creation of the initial individual list and its prioritising offsets the likelihood that respondents are influenced by the reactions of others and that the views of one or two individuals do not dominate. Each person has an equal opportunity to participate; there is also less need for respondent validation as the importance of each item is considered as part of the prioritising (MacPhail, 2001). In effect the respondents code their own data, reaching agreement on categories and coding them accordingly with less opportunity for the researcher to impose their own view. The specific questions we asked at the two time points (end Year 6 and beginning Year 7) varied slightly, but in each case we asked students to list the places they felt were most important and/or most enjoyable, and their opposites, and led the discussion from these initial lists. At time points 1 and 2, we focused on what we felt would be, or was, different about life in the new secondary school.

The map task

At the end of Year 7, we retained the same basic approach of eliciting individual responses and using these to prompt group discussion. We gave students maps of their schools and asked them to respond to a series of questions relating to their most and least preferred places around the schools, in terms of how safe they felt, and where was best to work or to socialise.

The school connectedness questionnaire

'School connectedness' is a concept that has been used in a variety of ways as an attempt to identify the psychological 'fit' of students to the school environment, encompassing elements such as health, security, social relations and self-esteem. The school connectedness questionnaire used in this project was modified slightly from that used by Goodenow (1993), which was a Likert-type scale seeking responses to a range of positively and negatively phrased statements relating to feelings of security, well-being and belonging. Goodenow developed a measure of youth connectedness to school, showing it to have high internal validity, with a Cronbach's Alpha score of 0.88. On this measure, the more nearly the score reaches 1.0, the more the items in the scale can be trusted to form a consistent measure of the construct under investigation. Goodenow devised the scale for use with 12 to 18 year olds, whereas the *Design Matters?* team used it with 11 to 13 year olds. A trial resulted in a slight reduction of the number of items, where we felt there was some degree of confusion among students about what an item meant. Our scale thus comprised 11 items. We also supplemented the five-point answering boxes (from 'Not at all True' to 'Completely True') with 'smiley' emojis showing a range of emotions connected with the relevant response. The questionnaire is reproduced in Appendix. This was our quantitative measure about students' feelings prior to, on entry to, and at the end of their first year of secondary school. This is considered more fully in Chapter 7.

The essay task

Students participants from 10 sample schools were asked to write an essay entitled 'My First Week at this School'. There was no prompting as to the importance of design or the building. We were keen to establish an understanding of their priorities and concerns on entering the school. This task was presented as a conventional activity in an English lesson, and they were marked by the English teachers accordingly. Copies of the essays were sent to the project team who analysed the content in order to ascertain how design and other aspects of school life featured in their early experience of the new school and identify students' priorities and concerns noted in the essays.

Observational methods used with students and teachers: 'third-person' approaches.

Formal classroom observations

(n = 55) lessons were observed using an observation schedule. These required us to make notes at five-minute intervals throughout each lesson about the nature of the teaching and learning activities being undertaken. We were particularly interested to ascertain whether such activities varied between types of schools, as well as within schools, and whether design features specifically intended to promote different ways of teaching learning, such as open learning areas and breakout spaces, were utilised in this way.

Photographs and informal observations

The research team took photographs of school buildings in use. In addition, we spent time undertaking informal observation tasks, such as sitting in cafeterias and walking round shared spaces such as corridors and playing fields during both lesson and break times. Our informal notes and impressions from these periods of observation, along with the photographic evidence, helped us form general impressions of each school.

Interviews

All interviews used a semi-structured format, comprising a series of lead questions that reflected our concerns at particular stages of the project. Interviews were either individual or in small groups, depending on context.

Documentary analysis

Documentary evidence was collected from school commissioners, architects and engineers to inform the understanding of the design brief and the commissioning and design processes. Documentary data comprised two main types: 1) records of the design-and-build process, and 2) statistics and other information relating to the school's performance and ethos.

1. We accessed records relating to the design-and-build process for each school. These came from a number of sources, within both the schools and the design teams.
2. We consulted school websites and brochures, and Ofsted reports. In addition, we obtained data concerning GCSE results, value-added scores, free school meals and special needs status from the Department of Education.

Data analysis

Our aim was to collect data in a variety of ways and from a variety of sources to inform debate about a broad area of educational concern: the effect of design on practice and experience of schooling. As with an ethnographic study, we were concerned to draw broad conclusions about the relation of design to practice and to experience, by drawing on a range of data that, taken together, form a strong data set. We wanted to undertake our analysis in such a way that broad areas of agreement would emerge. Thus we moved frequently between data sets, rather than focusing on each exercise as a discrete sub-project. There is considerable triangulation here; for example, the essay data reveal sets of student concerns very much in line with those arising from the NGTs and map tasks, thus giving each of these data sets greater explanatory power. The school connectedness questionnaire, which, as a quantitative instrument, was susceptible to a degree of inferential, as well as descriptive, statistical analysis.

In Chapter 6 we show how we juxtaposed qualitative and quantative data through time. Following Greene's (2007) discussion of crossover track analysis, we sought to construct narratives on the basis of the quantitative data and quantify the differences that emerged through the qualitative analysis. Our concern was to develop what Greene (2007, p. 156) refers to as a 'mixed methods way of thinking'. Bringing together data from a variety of perspectives, gathered through different approaches, allowed us to piece together a multifaceted account of the complexity of the perceptions and experiences of different school designs at a moment in time and as they changed through time.

All the interview data, including NGT and map task discussions, were transcribed. Content and thematic analysis was undertaken at the group or individual level (e.g. particular focus group or individual headteacher) and then at collective levels above that. For example, responses from individual NGT groups were compared with those from the same school then with those from neighbouring schools, then with the sample as a whole. This process enabled us to examine the 'trajectories' of students transitioning between different design and practice contexts. This work will be discussed further in Chapter 5.

References

Baars, S., den Brok, P., Krishnamurthy, S., Joore, J.P. and van Wesemael, P.J.V. (2018) Constructing a framework for the exploration of the relationship between the psychosocial and the physical learning environment, in W. Imms and M. Mahat (eds), Transitions Australasia: What is Needed to Help Teachers Better Utilize Space as One of Their Pedagogic Tools. Proceedings of an international symposium for graduate and early career researchers in Melbourne, Australia. www.iletc.com.au/publications/proceedings/.

Barrett, P., Zhang, Y., Davies, F. and Barrett, L. (2015) *Clever Classrooms: Summary Report of the HEAD Project*. Salford: University of Salford. http://ow.ly/Jz2vV

Botar, O.A.I. (2001) Notes towards a study of Jakob von Uexkull's reception in early twentieth-century artistic and architectural circles, *Semiotica*, 134(1/4): 593–7.

Bruner, J. (1986) *Actual Minds, Possible Worlds*. Cambridge, MA: Harvard University Press.

Burke, C. (2010). About looking: Vision, transformation and the education of the eye in discourses of school renewal part and present, *British Educational Research Journal*, 36(1): 65–82.

Burman, E. (2016) Assessing the operational performance of educational buildings against design expectations – A case study approach. Engineering Doctorate Dissertation. London: University College London.

Burman, E., Kimpian, J. and Mumovic, D. (2018) Building schools for the future: Lessons Learned from performance evaluations of five secondary schools and academies in England, *Frontiers in Built Environment*, 4: 1–16.

CABE (2005) *Picturing School Design. A Visual Guide to Secondary School Buildings and Their Surroundings Using the Design Quality Indicator for Schools*. London: Commission for Architecture and the Built Environment.

Chapple, M. and Murphy, R. (1996) The Nominal Group Technique: Extending the evaluation of students' teaching and learning experiences, *Assessment and Evaluation in Higher Education*, 21(2): 147–60.

Cleveland, B. and Fisher, K. (2014) The evaluation of physical learning environments: A critical review of the literature, *Learning Environments Research*, 17(1): 1–28.

Daniels, H. (1989) Visual displays as tacit relays of the structure of pedagogic practice, *British Journal of Sociology of Education*, 10(2): 123–40.

Daniels, H. (2001) *Vygotsky and Pedagogy*. London: Routledge.

Fish, S. (1980) *Is There a Text in this Class: The Authority of Interpretive Communities*. Cambridge, MA: Harvard University Press.

FNI (2011) *Educational Facilities Effectiveness Instrument*. Lutz, FL: Fielding Nair International. http://goodschooldesign.com/Default.aspx

Foxell, S. and Cooper, I. (2015) Closing the policy gaps, *Building Research & Information*, 43(4): 399–406.

Gagne, R. (1985) *The Conditions of Learning* (4th edn). New York: Holt, Rinehart and Winston.

Gearhart, M. and Newman, D. (1980) Learning to draw a picture: The social context of an individual activity, *Discourse Processes*, 3: 169–84.

Goodenow, C. (1993) The psychological sense of school membership among adolescents: Scale development and educational correlates, *Psychology in the Schools*, 30: 79–90.

Greene, J.C. (2007) Is mixed methods social inquiry a distinctive methodology? *Journal of Mixed Methods Research*, 2(1): 7–22.

Harré, R. (1983) *Personal Being: A Theory for Individual Psychology*. Oxford: Blackwell.

Harré, R. (1998) *The Singular Self*. London: Sage.

Harré, R. (2012) Positioning theory: Moral dimensions of social-cultural psychology, in J. Valsiner (ed.), The Oxford Handbook of Culture and Psychology. New York: Oxford University, pp. 191–206.

Hawkes, T. (1977; 2nd edn 2003) *Structuralism and Semiotics*. Abingdon: Routledge.

Hoffmeyer, J., Kull, K. and Sharov, A. (2017) *Biosemiotics*. New York: Springer.

Hirsch, E.D. (1987) *Cultural Literacy: What Every American Needs to Know*. Boston: Houghton Mifflin.

MacPhail, A. (2001) Nominal Group Technique: A useful method for working with young people, *British Educational Research Journal*, 27(2): 161–70.

Mumovic, D., Palmer, J., Davies, M., Orme, M., Ridley, I., Oreszczyn, T., Judd, C., Critchlow, R., Medina, H.A., Pilmoor, G., Pearson, C. and Way, P. (2009) Winter indoor air quality, thermal comfort and acoustic performance of newly built secondary schools in England, *Build. Environ.*, 44(7): 1466–77.

Palmer, J., Terry, N. and Armitage, P. (2016) *Building Performance Evaluation Programme*. Swindon: Innovate.

Punch, K. (2009) *Introduction to Research Methods in Education*. London: Sage.

Sanoff, H. (2001) *School Building Assessment Methods*. Washington, DC: National Clearinghouse for Educational Facilities.

Stables, A. (1998) Environmental literacy: Functional, cultural, critical. The case of the SCAA guidelines, *Environmental Education Research*, 4(2): 155–64.

Stables, A. (2006) *Living and Learning as Semiotic Engagement*. New York: Mellen.

Stables, A. (2009; first published 2003) School as imagined community in discursive space: A perspective on the school effectiveness debate, reprinted in Daniels, H., Lauder, H. and Porter, J. (eds), *Knowledge, Values and Educational Policy: A Critical Perspective*. London: Routledge, pp. 253–61.

Stables, A. (2012) *Be(com)ing Human: Semiosis and the Myth of Reason*. Rotterdam: Sense.

Stables, A. and Bishop, K. (2001) Weak and strong conceptions of environmental literacy: Implications for environmental education, *Environmental Education Research*, 7(1): 89–97.

Stables, A. and Semetsky, I. (2015) *Edusemiotics: Semiotic Philosophy as Educational Foundation*. London: Routledge.

Stables, A., Gellard, C. and Cox, S. (2018) Teachers' conceptions of students' 'ability': Creating the space for professional judgment, *Pedagogy, Culture and Society*. doi: 10.1080/14681366.2018.1491483

Stables, A., Murakami, K., McIntosh, S. and Martin, S. (2014) Conceptions of effort among students, teachers and parents within an English secondary school, *Research Papers in Education*, 29(5): 626–48.

Tarasti, E. (2000) *Existential Semiotics*. Bloomington, IN: Indiana University Press.

Williams, J. and Snipper, G. (1990) *Literacy and Bilingualism*. New York: Longman.

von Uexküll, J. (2010) *A Foray into the Worlds of Animals and Humans: With A Theory of Meaning*, trans. J.D. O'Neil. Minneapolis and London: University of Minnesota Press.

Wertsch, J.V. (1985) *The Social Mind*. Cambridge: Cambridge University Press.

Wittgenstein, L. (1967) *Philosophical Investigations*. Oxford: Oxford University Press.

Woolner, P., Hall, E., Higgins, S., McCaughy, C. and Wall, K. (2007) A sound foundation? What we know about the impact of environments on learning and the implications for Building Schools for the Future, *Oxford Review of Education*, 33: 1: 47–70.

Credit: BAM Construction Ltd

4
DESIGN AS A SOCIAL PRACTICE

Introduction

This chapter seeks to understand the ways in which the practices of school design produce educational spaces which mediate and shape the practices of teaching and learning when the building is occupied and in turn how the practices of teaching and learning shape the ways in which spaces become places for schooling.

Literature about the impact of physical environment suggests that school design can support student achievement (Woolner et al., 2007) but, as noted by OECD, current understanding of the use of spaces that are designed for the practices of schooling is still under researched:

> there is little existing research that focuses on how educational spaces are used as tools to facilitate the changing needs and demands of curriculum and pedagogy. However a growing body of literature highlights the need for learning environments to better respond to these needs and demands.
> *(Organisation for Economic Co-operation and Development, 2009, p. 17)*

Since OECD published a comprehensive literature review on the impact of school buildings in 2014, some progress has been made in more recent years. Two major themes emerge from this recent literature. The first dominant theme is the post-occupancy evaluation of the impact of physical learning environments on school improvement. OECD's (2014) literature review suggests that there is an overall lack of empirical evidence on the indirect effects that specific physical space qualities can have on learning and other outcomes. Since then a large proportion of the post-occupancy evaluation studies has been further developing this evidence base by using more rigorous research methods (Immes and Byers, 2016; Barrett et al., 2017; Lau et al., 2016), focusing on more specific types of outcomes (Brittin et al., 2017; Frerichs et al., 2015; Magzamen et al. 2017), and examining the possible mediators of the relation between school's physical environment and student achievement (Maxwell, 2016; Gilavand and Hosseinpour, 2016; Ariani and Mirdad, 2016). Some previously overlooked aspects were also explored. For example, a research team from the University of Nebraska – Lincoln (UNL) is currently working on establishing how the impacts of conditions in K – 12 school buildings on student achievement vary with different demographic factors (Lau et al., 2016).

Moreover, another stream of post-occupancy evaluation research focuses on the influence of school design on pedagogy in practice. These studies often examine the theoretical discourses of learning environments such as Innovative learning environments (ILE)/Mordern Learning Environments (MLE)/Flexible learning spaces (FLS) (OECD, 2013; Ministry of Education, 2017). Similar to the findings presented by OECD (2014), more effective teaching, increased use of diverse pedagogies and students' collaborative learning, alongside resistance to change can all be found in recent literature (Nambiar et al., 2017; Pantidi, 2016; Whyte et al., 2016; Deerness et al., 2018; Könings, Bovill and Woolner, 2017). It was suggested by Nambiar (2017) that teacher education is an important element if practitioners are to envisage the potential of Innovative Learning Environment spaces in reconstructing teaching and learning. Mäkelä (2018) suggests that maintaining the equilibrium between communality and individuality, comfort and health, and novelty and conventionality is crucial for the successful design of modern learning spaces. However, it is worth noting that findings from these studies still provide limited input to school improvement due to the fact that most of the studies are based on a small sample, often on single-school case studies. The schools' experiences might be contextual and unique to its own specific situation and community. In addition, insufficient attention has been given to the long-term effects on teaching and learning, with the notable exception of the study by Sigurðardóttir and Hjartarson (2016), where they collected data from a single school in Reykjavik after five years of occupation and found that, over time, teachers have been bending somewhat the initial design plan and leaning towards traditional teaching methods.

A second theme in recent literature is the increased focus on the school design process, especially participatory design process. There is emerging evidence suggesting that participation in design is more likely to motivate teachers to change practices and to refine their teaching. Some conceptual frameworks for participatory design were developed in recent years; for example, needs centred design (de Vrieze and Moll, 2016), substantive design principles and procedural design principles (Mäkelä and Helfenstein, 2016). Increasingly, research is also focusing on involving students as co-designers to ensure that they can play a key role in decision that could affect them (Wilson and Cotgrave, 2016; Can and Inalhan, 2017; Pearson and Howe, 2017). However, as suggested by Pearson and Howe (2017), pupils' views are at times in stark contrast to adult perceptions. Therefore, how to use children's views to facilitate change need further exploration.

Theoretical orientation

Sailer and Penn provide a set of proposals for the ways in which human activity, organisations and their social structures, and the production and use of spaces in buildings are intertwined:

> humans shape their buildings through design practice (social agency affecting spatial structure); humans shape their organisations through management practice (social agency affecting social structure); then buildings shape organisations (spatial agency affecting social structure); both organisations as well as buildings constrain agents in their behaviours (social structures and spatial structure-agency affecting social agency).
>
> *(Sailer and Penn, 2010)*

We found this set of arguments useful in developing our own theoretical orientation. We seek to build on their position through additional theorisation of these complex forms of relation. Our point of departure is Vygotsky's (1987) concept of mediation of human relations in and with the social world by cultural artefacts.

From this approach we have taken particular interest in the notion of tool or artefact mediation of engagement with an object in systems which change as contradictions are resolved. However, the emphasis on organisations and their social structures discussed by Sailer and Penn is not easily conceptualised in this framework alone. We argue that in order to fully understand mediation it is necessary to take into account ways in which activities are structured by their institutional context.

In the bid to explore the complex relationship between school designs and pedagogic practices, this section focuses on how designs of schools came to fruition. We focus on how the strategic educational vision for schools are developed and embedded in the final design. The *Design Matters?* project developed a methodology for systematically analysing the relationship of school space to the experiences of students, teachers and parents. It expands notions of post-occupancy evaluation (POE) research by exploring how the intentions of an educational vision which informed an initial school design, the intentions of the final building, and the intentions of those people who occupy that building interact in a way which influences experiences of the end-users. Crucially we looked at the social interactions that arose within a building as it was used over time. The analysis conducted allowed us to explore the process of design across different time periods. We found that motives can change depending on the aims and objectives at a particular point in time and that when the motives of different professional groupings differ at particular stages this can cause tensions. This analysis has provided us with a stage model which we will use for analysing how educational visions were developed and how these were translated into material spaces. The purpose of this is to provide a holistic understanding of how design processes impact on end-users' experiences of schooling.

Pedagogic post-occupancy evaluation

The field of POE has provided direction on how evidence can be gathered about the performance of educational facilities for over 50 years (Foxell and Cooper, 2015). Based on a critical review of the literature on the POE of physical learning environments, Cleveland and Fisher (2014) demonstrated that the majority of tools developed in various countries to evaluate school learning environments focus predominantly on the physical features of the physical environment itself. Cleveland and Fisher (2014) formed the conclusion that POEs on the social or human components of the learning environment in supporting pedagogical activities are in their infancy and require further development. In more recent years, the body of POE knowledge has been expanded by, for example, an inter-disciplinary theoretical framework developed by Baars et al. (2018) regarding the relationship between the psychosocial learning environment (PSLE) and the physical learning environment (PLE). However, further research is needed to test the preliminary PPR framework in an empirical setting to prove the validity, usability and reliability of the framework. Cleveland and Soccio (2015) developed an evaluation tool called the School Spaces Evaluation Instrument (SSEI), including Module 3 – Alignment of Pedagogy and Learning Environments. The SSEI tool was piloted in five schools with five principals, four assistant principals, 40 teachers and 222 students and was found to be an effective evaluation tool. The *Design Matters?* project developed a methodology for evaluating school buildings as a complex environment distinct from other building typologies that require a pedagogic understanding of the user's needs and the interactive relationship between the user and the physical space through time.

A significant proportion of the research on school design focuses on environmental issues, such as acoustics, lighting and temperature, typically uses quantitative methods, and often fails to explore how different environmental factors interact with users through time (e.g. Hygge, 2003; Galasiu and Veitch, 2006; Winterbottom and Wilkins, 2009; Shaughnessy et al., 2006). Furthermore, while there is a body of research which explores the relationship between buildings and pedagogies (e.g. Burke, 2010; Burke and Grosvenor, 2003, 2008), to our knowledge, there is a gap in research which explores the relationship of an educational vision to the creation, use and maintenance of physical space and, more importantly, how this relationship impacts on perceptions of educational experiences. This is surprising when one considers that an important aspect of the BSF process related to the development of a strategic educational vision and to ensuring this vision was transformed into a material design.

In this chapter, we present the findings from an analysis of the vision, design and construction processes in terms of key staging points. We focus on the personnel involved, and the decisions made, carried forward and abandoned at each stage. This overall research is aimed at exploring end-users' perceptions of school spaces; in order to understand teachers', students' and parents' perceptions of school space it is necessary to understand what that space was intended for and how it was intended to be used.

RIBA (2007–08, 2013) have developed a series of stages for the design and procurement process of construction projects and these are industry standards. Our analysis focuses on the stages of pedagogic development of the educational vision and how this vision was translated into a material design.

Methodology

Our methodology draws on previous experience of studying multi-agency working, albeit in the very different work places of child protection within and across child welfare services (Edwards et al., 2008). In these projects we identified post-Vygotskian Activity Theory as a particularly powerful tool for conceptualising and interrogating the sometimes fluid and rapidly shifting landscapes of these forms of developing professional work and the engagement with clients. At a very general level of description, activity theorists seek to analyse the development of consciousness within practical social activity. Their concern is with the psychological impacts of activity and the social conditions and systems which are produced in and through such activity.

An Activity Theory framework permits an analysis of the different 'object motives' (Leont'ev, 1978) evident in the practices of design, build and occupation; how they have emerged and are negotiated within and between the activities that arise in each form of professional practice (e.g. architecture, engineering); and which meaning systems prevail in shaping them.

Individuals acting with specific professional activities may behave in ways whose meanings are specific to a particular activity. From this perspective research seeks to identify and surface the contradictions that are embedded in everyday practices yet are often not visible to participants (Engeström, 1999). For example, there may be contradictions between motives such as predilections for particular forms of pedagogic practice, the aesthetics of particular aspects of design, acoustic performance and incentives to complete a project on time and within a budget.

In order to try to discuss innovation and improvement of specific forms of multi-professional activity, Engeström et al. (1997) develop a three-level notion of the developmental forms of epistemological subject – object – subject relations within a Vygotskian framework. They call these three levels 'co-ordination, cooperation and communication'. Within the general structure of coordination actors follow their scripted roles pursuing different goals (see Figure 4.1).

Within the general structure of cooperation actors focus on a shared problem. Within the confines of a script the actors attempt to both conceptualise and solve problems in ways which are negotiated and agreed (see Figure 4.2). The script itself is not questioned, that is, the tacitly assumed traditions and/or the given official rules of engagement with the problem are not challenged.

Engeström et al. (1997, p. 373) discuss reflective communication 'in which the actors focus on reconceptualising their own organisation and interaction in relation to their shared objects and goals (see Figure 4.3). This is reflection on action. Both the object and the script are reconceptualised, as is the interaction between the participants.

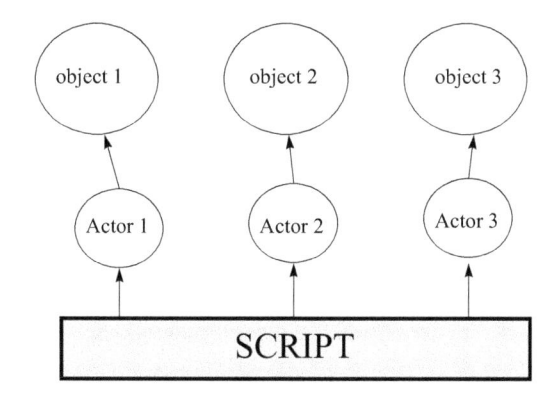

FIGURE 4.1 The general structure of coordination

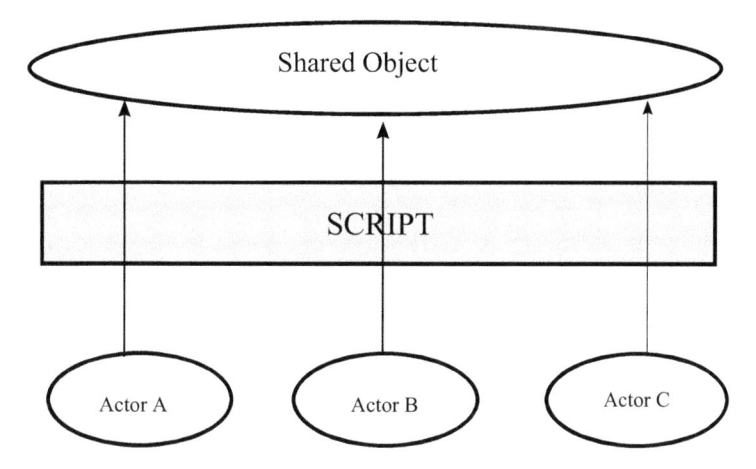

FIGURE 4.2 The general structure of cooperation

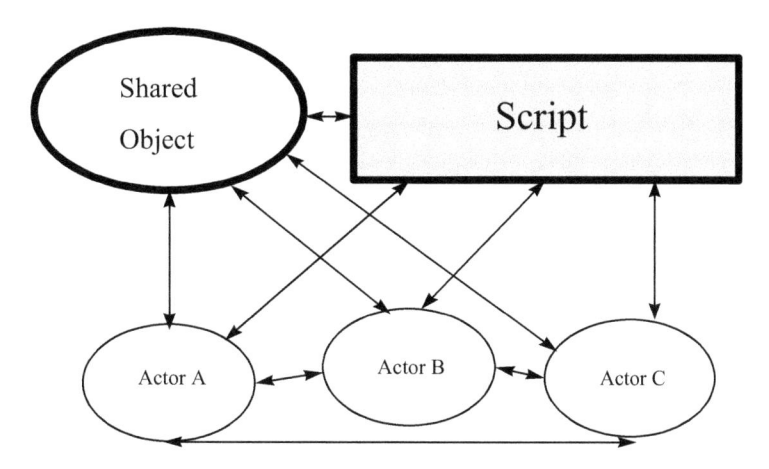

FIGURE 4.3 The general structure of communication

64 Design as a social practice

Implicit in this general structure of communication is a version of Vygotsky's (1978) concept of the Zone of Proximal Development (ZPD). That is the 'area that is beyond one's full comprehension and mastery, but that one is still able to fruitfully engage with, with the support of some tools, concepts and prompts from others' (Bazerman, 1997, p. 305). The description provided by Newman et al. of this form of activity in the classroom can be transposed to the actions of adults:

> The multiple points of view within a ZPD are not seen as a problem for analysis but rather the basis for a process of appropriation in which children's understandings can play a role in the functional system.
>
> *(Newman et al., 1989, p. 136)*

Within the complex networks of activity that we have studied there are two related forms of engagement. The first being between professional agencies and the second between professional agencies and clients. In the case of engagement across professional groups there has been increased emphasis on the way in which individual actors carry their own histories from the social positions that they take up in the division of labour that obtains within the activity. The idea of networks of activity within which contradictions and struggles take place in the definition of the motives and object of the activity calls for an analysis of power and control within developing activity systems.

One approach to theorising these matters has been developed by Engeström et al. (1999), who introduced the concept of knotworking to describe the 'construction of constantly changing combinations of people and artefacts over lengthy trajectories of time and widely distributed in space' (p. 345).

As regards the relationships between professional agencies and clients the term 'co-configuration' has been deployed: 'With the organisation of work under co-configuration, the customer becomes, in a sense, a real partner with the producer' (Victor and Boynton, 1998, p. 199).

As shown in Figure 4.4, Victor and Boynton (1998) argue that what craft workers know about products and processes rests in their personal intuition and experience about the customer, the product, the process and the use of their tools. When they invent solutions, they create *tacit knowledge* that is tightly coupled with experience, technique and tools. This is the kind of knowledge that teachers who regard themselves as 'intuitive' would develop and use.

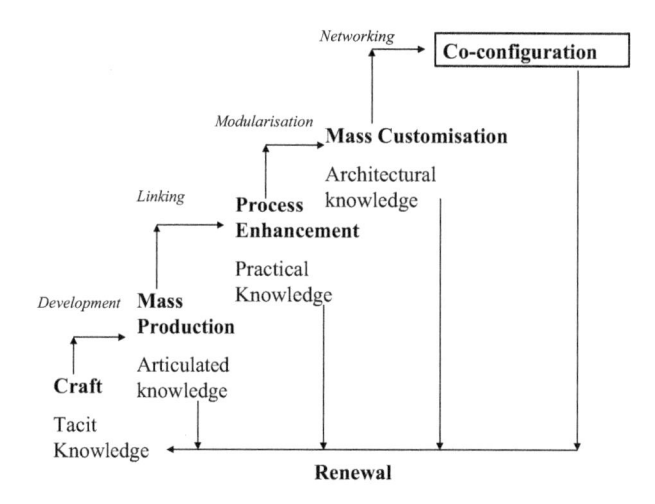

FIGURE 4.4 Historical forms of work (adapted from Victor and Boynton, 1998)

Through the articulation of the tacit knowledge of craft-work organisations may develop a machine-like system that mass produces on the basis of the knowledge it has 'mined' from craft work and reformulated as the 'best way to work'. This has been witnessed in attempts to codify and articulate 'best practice' in forms which are open to mass training and surveillance

Just as in the shift from craft to mass production, progress beyond mass production is created by learning. Mass production workers follow instructions yet also learn about work through observation, sensing and feeling the operations. They learn where the instructions are effective and where they are not. This learning leads to a new type of knowledge, *practical knowledge*. *Linking* is the transformation that bridges mass production by leveraging practical knowledge and creates the work that Victor and Boynton call *process enhancement*. It involves setting up a team system in which members focus on process improvement, which promotes the sharing of ideas within the team, and which fosters collaboration across teams and functions. This has been witnessed when initiatives such as the Literacy Strategy are implemented in schools.

The next move within their model involves the incorporation of the concept of *precision* alongside that of quality. The producer or service provider begins to try to identify precisely what it is that the client requires. This practice of *mass customisation*, progresses along a transformation path termed *modularisation*. The new knowledge generated by doing process enhancement work is leveraged and put into action as the organisation transforms its work to mass customisation. This transformation is based on *architectural knowledge*, the understanding of work required to make the transformation to mass customisation. Recent moves in the development and adaptation of curriculum and pedagogy in the 14–19 sector witness this kind of work.

Practices of mass customisation may be renewed when the available variety of options is exhausted. There may be the need to return to craft work in order to leverage out new information recognising that no universal formula can meet all client demands for precision. The crucial difference between this work and the move to co-configuration is that mass customisation tends to produce finished products or services whereas the emphasis of the next form of work is on the development of the product or service

Co-configuration typically also includes interdependency between multiple producers in a strategic alliance or other pattern of partnership which collaboratively creates and maintains a complex package which integrates products and services and has a long lifecycle. These already complex processes are rendered more challenging when there are several changes in client, multiple clients, or even confusion about the identity and nature of the client. It is, however, important to point out that some forms of design practice resemble craft work as discussed by Victor and Boynton (1998). In practices of craft work, knowledge often remains tacit. For example it might not always be possible for a creative artist or writer to articulate a plan or procedure for their way of working, rather it is something, like a habit, that has been acquired through practice. Our studies were of forms of work which involved different patterns of collaboration that resembled co-configuration and knotworking alongside aspects of craft work. The research that has been undertaken in construction and design has struggled to fully implement these ideas (Kerosuo, 2015). In our research into the complex processes of school design, we drew on the concepts of knotworking and co-configuration as points of departure in our analysis which sought to articulate different forms and typologies of collaborative effort. We found these ideas helpful in our attempts to describe and analyse the emergence of different forms of collaboration as design and construction initiatives progressed over time. We were particularly interested in the emergence of barriers and supports to progress in developing collaborative practice over time.

If reference is made to the Victor and Boynton model, then the origins of the work of designers can be understood in terms of craft work. As a result of financial and policy pressures, it would seem that the work of designers is being shifted into a form that resembles mass production with strict time limits and standardisation. Designers have, to some extent, responded by attempting to provide services of mass customisation for their clients but are severely constrained in the extent to which they can act in a responsive manner. Shifts in relations of control have resulted in the contractor becoming the lead agent in the process whereas in the past the client would have initiated the process with their vision which was realised by the designer.

Our research was based on a five-step, mixed methods design (Greene, 2008), collecting data through interview, surveys, observation and documentary analysis, capturing both first-person (subjective) and third-person (objective/intersubjective) perspectives over the phases of the project. The sample of schools consisted of 18 examples of different secondary designs including seven examples of 'traditional' schools used as comparators in our data analysis. Here we report findings from phase one of the *Design Matters?* project.

The research questions for this section of our research were:

1. How did the educational vision of a school come to fruition and how was this vision embedded in the final design?
2. How and why is the school used in the way that it is?

In this chapter, we report on the development of the methodology for pedagogic post-occupation of one school. An in-depth examination of the processes for vision, design, construction and occupation, we identified significant discontinuities at particular phases in relation to either the intended physical structure and what was actually built, and/or how space was intended to be used and how it was actually used in practice. What became very clear was that different motives were in play for different agencies at different moments in the process. For example, some agencies were driven by motives related to successful bidding for contracts at one moment and motives related to costs and completion on time at another. Many of these were in conflict with one another at critical times in the process which would lead to significant compromises for the built school environment.

Our aim was to interview one key member of staff from each organisation involved in the vision, design, construction and occupation processes. Interviews took place between November 2012 and September 2013 (n = 95). The research team also collected documentary evidence from school commissioners, architects and engineers to inform the understanding of the design brief and the commissioning and design processes. For the post-occupation analysis, the research team collected data (July 2013) from Y6 students (n = 371) to understand the students' expectations of the educational experience their new secondary school will afford and followed their transition into the sample schools. The research team spent one day at each sample school at three time points (September 2013, July 2014, July 2015) and carried out a set of data collection activities including an adapted school connectedness survey to identify the psychological 'fit' of students to the school environment, encompassing elements such as health, security, social relations and self-esteem (Goodenow, 1993), observations in class and break-time settings and student focus groups using a nominal group technique (NGT). This is a structured method for group brainstorming and establishing consensus that encourages contributions from all group members and fosters tolerance of conflicting ideas. Each person has an equal opportunity to participate; there is also less need for respondent validation as the importance of each item is considered as part of the prioritising (MacPhail, 2001). Data was collected to understand children's experiences of their new schools upon entry and through time how students and teachers utilise the spaces.

TABLE 4.1 Vision through to occupancy: stage model. Example: Locality C

Stage and year	Key organisations involved (BSF*)
1. Aims of BSF – Wave 3 (began 2005; started construction 2008; proposed finish 2014)	DfES
2. Local recontextualisation of 'Transformation' and a 'Strategic educational vision' 2005/2006	Council*; Community Members*; School*
3. Transforming the educational vision into a design 2006/2007	Council*; Project Management Company*; Concept Architects*; School*
4. Finding the Design-and-Build Team – invitation to tender; invitation to continue dialogue; invitation to submit the final bid This is the stage the Local Education Partnership (LEP) is developed 2007/2008	Council*; School*; PfS*; Consortium Team Leader; Contractors*; Delivery Architects*; Design Champion; Education Advisors*; Mechanical and Electrical Consultants; ICT provider; Facilities Management Services*; CABE Enablers
5. Design and build 2008–2010	Council; School*; PfS*; Consortium Team Leader; Contractors*; Education Advisors* Delivery Architects*; Mechanical and Electrical subcontractor; Acoustic Engineers*; ICT provider; Landscape Consultants; Facilities Management Services; CABE enablers
6. Process of occupation 2010 – date	Council*; School*; Delivery Architects*
7. Space to place 2010 – date	School* (Students*, Teachers*, Headteacher*, Parents*, Facilities managers*)

Data analysis

A semi-structured script was used for all the interviews. Interview data were transcribed and analysed. Two members of the research team conducted an inductive analysis of the interview transcripts to identify emerging themes within the data (Strauss and Corbin, 1998). The other two members conducted a deductive analysis relating answers directly to the research question. These analyses were then triangulated, allowing us to establish how the specific research question could be answered.

Using the information obtained from a review of the BSF literature and these interviews, we established seven key stages, focusing on the personnel involved and the decisions made, carried forward and abandoned at each stage and reasons why these decisions were made. Table 4.1 provides details of each stage and who was involved. Interviews were only conducted with personnel from organisations indicated by (*); these were the key people involved with developing the educational vision, translating this into a material design and how the design is used. The post-occupation analysis focused on the changing relationships between design and practice through time. The analysis sought to identify the priorities established through the practices of the schools and to consider why they emerged and how they impact on learning and teaching behaviours and outcomes.

Continuity and conflict in school design

In this section, we will describe in detail the development of the stage model in Table 4.1 to examine the vision, design and construction process of one school in south-east England. This analysis has provided us with a stage model which we used for analysing how educational visions were developed and how these were translated into material spaces in 10 newly designed schools. The purpose of this is to provide a holistic understanding of how design processes impact on end-users' experiences of schooling.

Stage 1: Building Schools for the Future – contextualising the school's design and build process

As the target school was procured as part of BSF Wave 3, a starting point was to explore the BSF literature in a bid to understand what the motives for BSF were at the time School C4 was procured. While it is not within the remit of this chapter to provide a full overview of BSF (see Mahony et al., 2011 for a detailed account), in this section we provide an overview of why the BSF programme emerged, an insight into the BSF programme at the time School C4 was procured and an overview of the current situation relating to school building programmes.

School-building programmes

With growing concern about attainment levels of children in UK state education, and after years of neglect in relation to school buildings, in 1992 the introduction of the Private Finance Initiative (PFI) resulted in increased investment in school buildings; concerns that these were not fit for purpose (e.g. Ofsted, 1999–2000) led to the development of new privately financed state school buildings. These were funded through a combination of state finance and the PFI. By 2002, although over 500 primary and secondary schools were part of PFI deals signed into procurement, only approximately 25 new-build PFI schools had been completed (Audit Commission, 2003).

To our knowledge, there were no guidelines relating educational visions to school design at this time and, perhaps because of this, a commissioner from CABE described these schools as 'little better than agricultural sheds with windows' (UNISON, 2003, p. 9). In a bid to 'seek and encourage innovation through school building design' (CABE, 2007, p. 8), the Labour government introduced the Building Schools for the Future Programme (BSF) (DfES, 2003).

BSF was aimed at rebuilding and refurbishing all 3,500 secondary schools in England between the years 2005 and 2020 and 'the use of high quality ICT in these new bespoke schools was seen as a means of transforming the learning experience of students and raising attainment' (House of Commons: Education and Skills Committee, 2007, p. 12). The programme encouraged private sector participation in planning, financing, refurbishment and delivery of educational services via joint ventures with local authorities; private finance initiatives (PFIs) were the preferred way of paying for BSF (Mahony et al., 2011).

The BSF agenda was directed towards these newly designed schools being educationally and socially transformational thus 'playing a key role in the Labour Governments educational and social policy which was aimed at making a more equitable society' (Mandelson, 1997, p. 7). The programme was seen by some as a 'deliberate attempt to drive reform in the organisation of schooling, teaching and learning through the delivery of innovative school buildings' (Education and Skills Committee, 2007, p. 91). BSF was aimed at 'expanding the infrastructure of social and educational provision' (Mahony and Hextall, 2013, p. 854) to include the wider community and therefore 'principles of transformation, redistribution, regeneration and participation were all intended to play their part in the BSF initiative' (p. 854). It was apparent that the Labour government saw new school buildings as crucial to educational and social reform (Kraftl, 2011) and that Labour believed that capital investment was vital to quality of learning (DfES, 2003, p. 4).

There was no vision for PFI, but the vision had arrived for BSF. They'd [the Council] made a serious mistake with the original PFI programme where they'd rebuilt 6 new schools from the old model, old new schools effectively.

(Secondary Transformation Team interview)

Under BSF, schools were required to produce strategic educational visions which would translate into material spaces in the eventual design (PfS, 2008, p. 15). It was recommended that teachers, parents and pupils should all be central to the development of this vision (PriceWaterhouseCoopers and DCSF, 2004). Following their review of BSF, the Education and Skills Committee (2007, p. 3) suggested 'there is a strong argument that the initial "visioning" stage should be lengthened'. This suggestion was based on the claim that 'people involved with BSF, particularly at the school level, did not have enough time to think about what they wanted for their new school' (p. 3) and particularly 'issues about how secondary education should be provided in their area' (p. 62). The review also suggested a 'clear statement of the national ambitions for twenty-first century education was required to provide guidance for this vision developing process' (p. 62).

While more guidance was provided by Partnership for Schools (PfS) – in 2006 all schools/academies procured from Wave 5 onwards were required to produce a 'Strategy for Change' (SfC) document on the visioning process – this guidance was only available for Wave 5 onwards, and while it provided schools, academies and local authorities with assistance on how to produce a vision or SfC document outlining how the transformed school would meet the needs of the educational experience of students (Chidgey, 2009), little was available in relation to ensuring a continuous process in which the development of this vision was effectively translated into a material design.

CABE (2004, 2007) did recommend that in order to achieve a good design it was not only essential that a clear vision was developed, but also recommended that a client design advisor be allocated to each project to:

help schools to translate their vision into a brief, and help them to challenge design proposals that fall short of their aspirations and evaluate designs as an 'expert client'.

(CABE, 2007, p. 12)

The problem with the guidance is that they provide no insight into multi-agency working and the challenges of this process.

With large amounts of public expenditure earmarked for the BSF programme, a number of reports began to emerge questioning the success of BSF (e.g. Shaoul et al., 2010; Kraftl, 2011; NAO, 2009; PAC, 2009). These were predominantly related to the procurement process which was considered to be far too expensive, overly complicated and wasteful (NAO, 2009; PAC, 2009) and to the delivery process, which was considered to be 'extremely complex' (Mahony et al., 2011, p. 348).

The current situation

In July 2010, the James Review was launched by the Secretary of State. Its aim was to examine how education capital in England was spent and to make recommendations on the future delivery models for capital investments for 2011/12 onwards. The review suggested it was a mistake to start a building project with a sum of money rather than a specification as this resulted in designs which were 'far too bespoke'; instead, building programmes should start with a focus on the output wanted, for example, size/scale of building (James, 2011, p. 5). Concerns were also raised about the lack of expertise on the client side (the local authorities and headteachers); this reportedly resulted in lost opportunities because 'good' ideas were not being shared across sites. Recommendations were made for School Building Programmes to become centralised, saving money and time and also ensuring lessons were learned from one project to another. In 2012 Michael Gove announced the 'Priority Schools Building Programme' (PSBP; Livesey, 2012), which introduced the idea of standardised school design.

The main aim of the PSBP, in line with proposals made in the James Review, is to ensure the procurement process is highly centralised. The rhetoric behind the programme argues that while there is clearly a need for new school buildings, there is simply not the money available that was spent during the BSF era (Elledge, 2012). The proposed £2.69 billion PSBP is to be funded mainly by PFI and is to be used to cover the equivalent of building 100 secondary schools (Livesey, 2012). The programme has four objectives:

> Ensure schools in the poorest conditions are prioritised for rebuilding (BSF aimed to rebuild schools in the most deprived areas of the country);
> Build more for less by drawing on learning from BSF and other school projects;
> Improve PFI model by drawing on learning and experience from recent BSF/PFI deals;
> Use a centralised procurement model to drive efficiencies in procurement time and ensure a quality solution is obtained at the lowest cost.
>
> *(Livesey, 2012, pp. 7–10)*

Other than suggesting this approach will provide the benefit of 'preventing poor educational outcomes due to substandard building conditions', unlike with BSF there is no mention of 'transformation' and no mention of designing schools which provide an adaptable/flexible environment for changing pedagogical needs. To date nothing is written about strategic educational visions in the PSBP literature; the development of an educational vision for new school designs is apparently not deemed necessary, perhaps because its elimination cuts out the early stages of the procurement process and saves money (see Livesey, 2012; DfE, 2011).

The dominant motive for building schools has changed. Whereas BSF was a vehicle for driving social and educational reform (Education and Skills Committee, 2007), under the coalition government the motive behind standardised school designs is to address concerns for the state of the country's school buildings and to meet demands for new school places as cost-effectively as possible.

While BSF required schools to develop a strategic educational vision, and even though guidance was provided for how to develop this vision, no clear understanding of what was meant by transformation was provided by BSF (James, 2011, p. 12) and, more importantly, to our knowledge little was understood about the actual process of how an educational vision translated into a material design, nor subsequently the impact of this on the educational experiences of end-users. We argue that in order for recommendations to be made in relation to school design, it is first necessary to develop an understanding of the issues of continuity and conflict in the design and build process, not just for BSF schools but for all school build programmes, and to explore the extent to which any issues observed impact the final build and the experiences of end-users.

Stage 2: the recontexualisation stage

An examination of the vision documents of all the five schools involved in the larger project demonstrates how those parties involved in the development of an educational vision attempted this by interpreting government policy on building programmes and tried to develop these ideas in line with local contextual needs. We say 'attempt' because it soon became apparent that local contextual needs were not always taken into consideration. We refer to Stage 2 as the 'Recontextualisation' stage.

Stage 2 includes an overview of the school and an overview of the educational vision.

School C4

School C4 was built as part of Wave 3 of BSF with design and procurement beginning in 2006. It is a local authority-maintained school located in the south-east England. Built in 2010, it has the capacity for 890 students; currently there are 751 on roll. The number of students eligible for free school meals is above the national average as is the number of students identified as having Special Educational Needs. School C4 is located in one of the most deprived communities in the country.

Following a successful application to DfES in 2004, School C4 was selected to be one of the council's four sample schools to participate in Wave 3 of the BSF programme. It was developed as an exemplar design (Figure 4.5) tailored to meet local needs and aspirations (DfES, 2004). As David Miliband MP, Minister of State for School Standards, stated:

> To transform our secondary schools . . . High quality, modern school buildings, with the latest integrated ICT systems, will help to raise standards and will play a crucial part in our ambitious programme of educational reform.
>
> *(Schools for the Future, 2004)*

In 2002, the county council initiated a review of secondary provision and, in partnership with schools, developed a strategy to reshape the county's educational landscape influenced by BSF's 'Transformation'

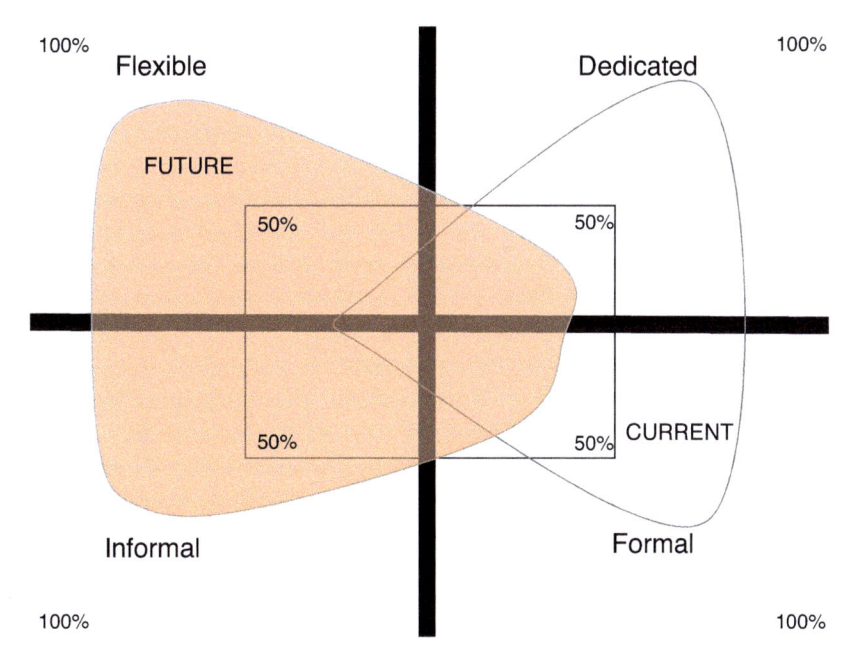

FIGURE 4.5 Educational model for new learning spaces

Source: Nurturing Autonomous and Creative Learners: the XX Secondary Strategy – Phase 2, BSF Documentation County Council, 2005.

Agenda (DfES, 2003). In a bid to develop the council's secondary strategy, and ensure transformation occurred, they recruited a Secondary Transformation Team (STT) of ex-heads and educationalists. The STT set out to get the county's schools to work collaboratively towards improving standards. This work became the development for the Schools Improvement Partner Programme for the Council (2006): 'We had to understand what transformation meant to us?' (STT interview).

From the onset, there was a clear policy priority to use BSF to deliver educational transformation but there was no evidence of a 'coherent definition' of this central objective. (James, 2011, p. 12). Furthermore, the lack of guidance led to variations in the programme, process, design and outcome as each of the bodies involved had to implement their own interpretation of 'educational transformation' (NAO, 2009, p. 6; James, 2011, p. 13).

The educational vision

> It wasn't about buildings, it was about transformation. This is about education and the principles that you want to operate should work in the middle of a field without a building.
>
> *(STT Interview)*

The STT recognised they needed to address local contextual needs and involved headteachers from each of the eight schools involved in the county's BSF programme (Wave 3 and Wave 4) in developing their own educational vision. Numerous workshops were held with members of the schools, the STT and the council. These professional groups worked together to develop a new strategy which would provide the basis for the educational vision for the council's BSF programme and ensure transformation of the council's educational ambitions. According to a member of the STT, the council's dominant motive for this transformation was to increase academic attainment because they were underperforming in relation to the desired for five A★ – C grades at GCSE level. This supported the Labour government's drive to improve the attainment of young people at a time when reports suggested the UK was falling behind.

Four areas were identified as core elements for the county's BSF programme:

1. Transforming the organisation of learning in schools.
2. Developing capacity and structures by extending collaborative partnerships between schools by developing clusters and transforming them into Education Improvement Partnerships.
3. Placing schools firmly at the heart of their local community.
4. Designing and developing the learning infrastructure in collaboration with the sample schools.

These all reflect the Labour government's drive for social and educational reform embedded in the key priorities of the BSF literature. Innovation and personalised learning were also emerging BSF themes that encouraged schools and design teams to transform the traditional school environment. As CABE stated, 'the emphasis of this programme is on transforming educational attainment and standards through innovative school buildings' (CABE, 2004), and yet the James Review highlighted that there was very little evidence that a school building that goes beyond fit for purpose had the potential to drive educational transformation at this time (James, 2011, p. 13).

The council published their Template for Schools of the Future document in 2005 which established key strategic approaches on innovation, flexible approaches to learning, personalisation and community participation. The template also highlighted that collaboration between the users and designers is a key component to success. The council commissioned a group of leading consultants to develop the template's early design proposals to be used as starting points for the BSF designs, but none of those consultants were involved in the following stages of design development for the council. There was also little evidence that subsequent members of the design team used the council's template to inform or expedite the design or consultation process with the individual schools.

At the level of policy, the Department of Education and Skills also appointed 11 leading design teams in 2003 to develop exemplar designs (DfES, 2004) for schools to be used as innovative BSF models. In practice, we also found little evidence that these designs were actively used by design teams or stakeholders.

The council's template also developed an educational model for students to be taught in 'learning clusters'. The cluster concept was based on the principle of 'schools within a school'. The structure offers small-scale educational nuclei which support a strong pastoral-care system and allow each school to choose its own grouping of specialist areas depending upon needs (KCC, 2005). The strategy was for students to be taught in a range of different size learning groups (from small group work up to groups of 240) to support the move towards autonomous learning.

It was apparent that the STT and the council had a strong commitment towards ensuring this new educational vision was championed by each school and their design teams. Each school, led by the head-teacher, worked with the Sample Schools design team and the council to develop their own learning clusters of specialist areas within the council's concept set out in their template.

The rhetoric of the Secondary Transformation Team and the council indicated that the council's secondary strategy was a distinctive pedagogical programme springing from the specific and urgent needs of the local educational landscape. While this was no doubt the case, a number of the insights and innovations were appropriated from other parts of the UK and abroad in quite different circumstances from the contexts in which they were developed and implemented. The BSF literature actively encouraged schools and designs teams to seek inspiration and visit innovative schools in the UK and abroad (CABE, 2004):

> Building Schools for the Future will drive innovation and transformation in education: each wave of Building Schools for the Future will comprise projects where innovation can have greatest impact on standards.
>
> *(DfES, 2003)*

A key part of the visioning process at the council was a trip to the United States in 2005 which involved 93 secondary school heads on a fact-finding trip to survey selected American charter schools to gather data on programmes, pedagogical initiatives and exchange ideas with innovative US educators:

> We took 93 out of 105. We also took officers and advisors as well. So we had 25 in California, 25 in New York, 25 in Boston, 25 in Seattle and then XXX sponsored a two day conference in Seattle for the whole lot.
>
> *(STT interview)*

While this strategy to look further afield accorded with BSF guidelines, evidence that something works in one context does not necessarily mean the same will be relevant in another context. As pointed out by Cartwright (2007), evidence-based policies should highlight the importance of 'evidence that deals with both soundness and relevance and that are at the same time both principled and practicable' (p. 17). Interestingly, we have found a similar situation in the other schools involved in our project; all have gone beyond their local context when developing their visioning process and have implemented numerous ideas with little evidence that these ideas can work in their local educational context.

The council's rationale for change was grounded in an intention to replace traditional teaching methods which focused on a 'chalk and talk' model of education in which students were passive recipients of this knowledge to one in which students became actively engaged with the learning experience through inquiry-based learning:

> These buildings have been designed in a way that makes it almost impossible for them to go back to square one. That was our guiding principle.
>
> *(STT interview)*

The council's strategy for transformation was to provide the vision for new ways of delivering secondary education in the county. It followed the core elements of the BSF programme, which promoted personalised learning in curriculum content, assessment, learning style and different forms of learning. Models were developed for a diverse range of open-plan learning spaces which illustrated features of adaptable environments, addressing acoustic and ICT needs for learning in a variety of group sizes, from individual learning to large group sessions. These design templates would serve as models for how personalised learning could be delivered in the Council's new BSF schools:

> There was such a strong commitment from the STT coming from [council], I do think [they had] the vision for the schools as well, and for the philosophy behind education for [council] and I just loved the flexibility to be creative about it.
>
> *(Second headteacher, School 3)*

The strong drive for innovation from the Council and DfES pushed through the new models on personalisation and demanded commitment from the heads of the BSF schools involved.

From our interviews, there were significant concerns from the schools and the design teams about whether the new open-plan learning models can work in practice but the council showed strong commitment at this stage to support the schools on the delivery of their educational vision.

Stage 3: transforming vision into design

Stage 3 refers to the process of taking the educational vision to the conceptual design stage. In March 2006, the council, in collaboration with the school communities, began work on the BSF Sample Schools project to help develop each school's educational vision for the outline business case. The council commissioned a project management company, who in turn commissioned a firm of leading architects in the field of educational buildings for the five sample schools. The Sample Schools project aimed to translate the council's vision of 'autonomous and creative learners' (CC BSF template, 2005) into five preliminary school designs. These were conceived to serve as a design brief for the procurement process and act as a springboard for the bidding teams. However, from our interviews and the NAO review in 2009, 'the bidders felt these initial designs were not useful and chose to start again when producing their bid' (NAO, 2009, p. 47).

The Sample Schools architects were commissioned to translate the council's BSF educational vision into preliminary designs (equivalent to RIBA Stage B/C) within an eight-week time frame. They had 25 workshops with each of the five sample schools in two towns within the council. These workshops were aimed at developing each individual school's pedagogical vision and translating these ideas into a new 'accommodation schedule' within the BB98 funding envelope. As a member of the Sample Schools architectural team stated:

> The educational vision was to create groupings around subject clusters so we introduced a technology zone, an art zone, a PE zone, and general learning areas [. . .] Each cluster was physically

organised around a breakout space supported by general learning areas, cluster had a specific specialism, the science zones had labs associated with it, the technology zone had technology areas and so on. Overlaid over the whole was the virtual pastoral organisation of the school.

The Sample Schools design team led by the council worked in collaboration with the five schools to translate the council's educational vision into preliminary exemplar designs (Figure 4.6) which would become the innovative models for future BSF schools in the council. Our interviewees all expressed positive feedback on this stage of the process. A close working relationship was forged between the council, the school and the architects, which would subsequently change as the BSF procurement process took over. This part of the process was about extensive consultation and about creative engagement with the community. The architects, the school and the council mobilised a unified, collaborative approach towards an early conceptual design which they agreed met the requirements of their educational vision. This was the last stage in which the council's educational vision was central to the design process. The continuity of leadership by council ensured that the new educational vision set out in Stage 2 was embedded in the early development of the school designs in Stage 3. The major conflict at this stage was due to the complexity and bureaucratic requirements of the BSF procurement process. From the onset of Stage 2, the Sample Schools architects knew that they would not be involved with the delivery of the new BSF schools. While the council and the schools benefited from a close engagement with the Stage 2 design process, which was a key part of the BSF agenda on community transformation, it is appropriate to question the usefulness of these early designs in practice. The nature of the design process meant it was difficult for bidders to assess the affordability, the technical performance and the design priorities of these initial designs. Without continuity and a strong framework for the accountability of the design work, the

FIGURE 4.6 School C4 concept exterior

Credit: County Council

bidders 'chose to start again when producing their bids' (NAO, 2009, p. 47). A total of £12 million was spent on consultants' fees by council for the Samples Schools project to Stage 3:

> The process was too complex and we, as a transformation team were often saying this is nonsensical . . . we know what we want . . . schools know what they want . . . in our private discussions in the transformation team we would say well XXX (Sample Schools architects) have now come up with all the designs we want, let's do it.
>
> *(STT interview)*

Stage 4: finding the design-and-build team

Stage 4 assesses the competitive bidding process of the BSF procurement for the schools. The council was one of the early adopters of the BSF competitive dialogue process. The purpose of the process was to find a consortium to fund, construct, and maintain the schools.

During the competitive dialogue process, the council selected two bidders to design seven schools from 12 initial bidders, before selecting its preferred bidder to further develop the designs for four schools before financial close. The bidders reported that the onerous process placed 'unsustainable strain' on their bidding capacity (NAO, 2009, p. 47).

The data from our interviews showed Stage 4 produced a problematic discontinuity for the educational considerations of the proposed designs. The educational vision was no longer the key priority in the development of the project for the design team and, while it was apparent the school community felt fully involved in the development of the educational vision in Stages 2 and 3, the subsequent BSF procurement process disengaged the end-users from the designers of the schools. This occurred despite clear guidance in CABE and BSF literature on the importance of broad consultation (CABE, 2004). Although Den Besten et al. (2008) recommends that 'clear, national, documentary guidance about when, where and how pupils should be positioned with school design' is required to promote non-tokenistic pupil participation in UK school design.

The council and the schools had to invest intensive periods of time into the competitive dialogue process but the participants we interviewed felt they were being asked the same questions they answered in Stages 2 and 3. The outcome of the process was also 'disappointing' for the schools, as the consortium that produced their preferred designs did not win the bid. (The selection process for BSF PFI projects weighs design, financial, legal and facilities management (FM) issues and therefore it is not always possible to select the bidder with the preferred designs.)

The delivery architects for the preferred bidder expressed their experience of a discontinuity in the design process, resulting from conflict between their professional role to understand and design for the needs of the end-users and the formal routes of communication during the competitive bidding process:

> We'd had a couple of meetings with the school prior to winning the bid. I imagine the discussions in those meetings were maybe slightly dysfunctional as they were developing three designs with three different consortia in parallel. It must have been hard to focus with three different designs in front of you. Probably a bit like being a juror at an architecture crit – something I'm sure none of them had much experience in!
>
> *(Delivery architect interview)*

The Sample Schools projects were presented to the bidding consortia and their design teams but there was no formal handover of information between the Sample Schools architects and the selected delivery architects:

> We won the bid with a design which wasn't properly checked for affordability. Corners were probably cut because of the sheer amount of materials that had to be produced for the bid package. Also, to increase the consortium's chances of winning a very large chunk of work, there was no doubt pressure on the whole bid team to produce, for this small sample of schools, an attractive rather than an affordable design. For example, it turned out that the roof lights were too costly, but they had to be built because they were clearly shown in the bid drawings. As a result other building elements that weren't so clearly drawn suffered, such as acoustics and mechanical services.
>
> *(Delivery architect interview)*

The school had just completed an intensive eight weeks of consultation with the Sample Schools architects during Stage 3, and the school felt they had fully collaborated on designs that met their educational visions. However, the BSF procurement process changed the consultation radically at this point and the school was then distanced from the bidding and subsequent delivery process. Our interviewees expressed concern that any open engagement between the school and the design team was viewed negatively, as opportunities for the school to change their brief could affect the cost of the project. The school felt disengaged from the design process from Stage 4. All subsequent engagement was viewed by the school and the design team as no more than a tokenistic attempt at school participation.

By the time the school was presented with the final design, they were no longer in a position to engage in active consultation as the design was fixed and the proposal had already been submitted for planning:

> I joined the project after we had won the bid, and to me the whole process of developing the design with the school after that seemed wrong. I expected open, informed and dynamic discussions, held within openly stated constraints of affordability. Instead we got only very limited contact with the school, carefully choreographed by the main contractor so that we didn't say a word out of place. The school probably felt they should have had a lot more say in the development of the design than they ended up having.
>
> *(Delivery architect)*

All communication between the bidding architects and school was formally exchanged in writing through the project bid director during the competitive process.

The conflicts explored in Stage 4 indicate a substantial disconnect between the aims of the BSF programme and the organisational structures and protocols of the building process for School C4. The users felt alienated from the process and the design architect believed that opportunities for constructive engagement had been missed, leading to a suboptimal design that failed to deliver on the promise of the BSF programme.

Stage 5: design and build

Stage 5 describes the construction phase of the project and how the decisions made impacted the learning environment.

78 Design as a social practice

The council's educational vision set out to 'nurture autonomous and creative learners'. The delivery architects responded by developing a concept (Figures 4.7 and 4.8) that 'will allow the users to respond to new learning styles, adapt to changing pedagogy, and accommodate internal reconfiguration' (Delivery Architects, ITCD bid document, 2008, p. 3) within an adaptable architectural framework. The delivery architects designed innovative open-plan spaces that aimed to facilitate different styles of learning but the usability of the built open-plan spaces were undermined by the contractor to meet budgetary requirements during this stage of the project. The cost pressures resulted in downgraded Mechanical and Electrical services, acoustics and FF and E (Furniture, Fittings and Equipment) specifications which meant the open-plan learning spaces, central to the educational vision to 'nurture autonomous and creative learners', could no longer be delivered at the proper level of quality.

The decisions made at this stage were critical to whether the initial educational vision of the school could be delivered. Different team members felt frustrated by the conflict between their contractual obligations to their client (the contractor) and their professional obligations to the end-users (the school). The delivery architects stressed that there was a notable absence of the role of a quality and design compliance monitor working for the council and the school during this stage of the project to ensure the built design would provide educational spaces that could deliver the educational vision set out at the inception of the project:

> It was identified in the bid document that the compliance of the design with BB93 standards for speech intelligibility could only be robustly assessed by computer modelling. Therefore, as part

FIGURE 4.7 School C4 learning zone sectional perspective

FIGURE 4.8 School C4 concept plan
Credit: Feilden Clegg Bradley Studios

of the BB93 process, computer modelling of the open-plan spaces was carried out, which confirmed that the BB93 speech intelligibility targets would not be met. At this point in the process, in accordance with BB93, the client [council] was asked to confirm that although when activity noise levels were at a normal level, the building design would not comply with the BB93 criteria . . . for 'educational reasons' the design was acceptable to them.

(Acoustic consultant)

Guidance from Building Bulletin 93 (2003) specified that 'Open-plan spaces require extra specification as they are significantly more complex acoustic spaces.' Guidance also specified performance standards for speech intelligibility in open-plan spaces and emphasised the importance of 'ensuring that open-plan spaces in schools are only built when suited to activity plan and layout'. Not only were the acoustic specifications at School C4 not enhanced, they were drastically reduced to below-standard specifications for conventional classrooms. Approved layouts for Furniture, Fittings and Equipment for the open-plan learning spaces at School C4 were also downgraded to low-quality specifications that can no longer facilitate flexible styles of learning or support teachers in delivering new styles of pedagogies set out in Stage 2:

> It means that management of teaching activities and control of activity noise throughout the school is key to achieving a learning environment aligned with council's educational vision.
>
> *(Acoustic consultant)*

The council suffered a lack of continuity and capacity within the STT team, with key members of staff leaving during the construction of the school. The ambitious educational vision required continuity of leadership and support for the schools to deliver a built environment that can support the challenges of new teaching and learning methods. Key cost decisions at this stage required to be assessed against how it would impact the delivery of the school's and the council's educational vision:

> The organisation at the hub of the Client, Designers and Consultants is the Contractor and one of their prime drivers is to control costs so that they can deliver the building in a timely manner and generate a profit. It would be nice (for the designers) if the prime driver was to have an excellently designed building, but the excellence of design often took second place to the priorities of controlling costs and therefore generating profit.
>
> *(Acoustic consultant)*

Stage 5 of the project describes the final design and construction phase. The contractor is the leading member of the delivery team for this phase and is under high commercial pressures to deliver the project on time and under budget to the agreed specifications of the design at financial close. The cost pressures in this construction stage have resulted in key decisions that not only impact on the present but also future flexibility and adaptability for the end-users. It can be argued that this phase of the project is one of the most critical phases for monitoring quality and compliance to the aims of the educational vision of the project. Decisions made at this stage, often under cost pressure, can fundamentally compromise the ability for the built design to deliver the educational vision of the school:

> Not enough detail was included in the design at both bid submission stage and financial close stage. This meant that the client was very exposed to substandard quality creeping into the design as we detailed it. The main contractor was unwilling to share with us the cost plan during detailed design development, nor use our design expertise in cost-cutting discussions. In the end, they had ignored many of our drawings and specifications and built the way they wanted to, to the detriment of quality and performance.
>
> *(Delivery architect)*

Stage 6: process of occupation, and Stage 7: space to place

Stage 6 refers to the process of the end-user coming to occupy the school building and Stage 7 refers to the process of the end-user both shaping and being shaped by the building. Despite the published BSF literature and guidance documents from CABE and RIBA, these stages of design projects remain under-researched. The current work on post-occupancy evaluation (POE) tends to be narrowly focused on environmental factors and quantitative data analysis (e.g. Hygge, 2003; Galasiu and Veitch, 2006; Winterbottom and Wilkins, 2009; Shaughnessy et al., 2006). The next phase of our research is aimed at exploring these stages.

Discussion

> What we were interested in was education, not buildings, but at the same time you can mess up the quality of education you provide for a child by giving it a lousy building.
>
> *(STT interview)*

The project analysis allowed us to develop a timeline of events in the design and procurement process of this case study school. These were placed into the specific stages and the themes which emerged for each of these stages provided an interesting insight into where and why key decisions relating to an educational vision were made; where tensions and trade-offs arose; and how this influenced the extent to which the educational vision was translated into a material design. Table 4.2 provides a summary of these main points for each stage:

In Stage 1, we explored the BSF agenda. BSF was a vehicle for driving the Labour government's educational and social policy (Mandelson, 1997, p. 7); the dominant motive of the programme was to push the government's political agenda. A secondary motive was to find ways of funding BSF, thus the encouraged use of PFIs. Nevertheless, it was the public political rhetoric for transformation which underpinned the educational vision developed for School C4. While there were numerous guidelines for BSF and for the development of a strategic educational vision, guidance on the actual process of design-and-build (e.g. CABE, 2004) offered no insight into how and why potential conflicts might occur when engaged in multi-agency working, how these could potentially influence the (dis)continuity of a project, and the implications of these on the final build.

It was apparent that during Stages 2 and 3 some of the professional groupings had different motives. However, the aim to create an environment which met local needs and transformed teaching and learning was eventually shared between the council, the school, the concept architects and the STT. There was also a drive to innovate and create new schools that could not revert back to traditional teaching methods. The motive for the school was to increase academic attainment and to improve inadequate facilities for students and staff. When the concept architects came on board they adopted the same aims and, although they had different motives, as with the school, this did not appear to cause any tension between the different professional groupings. To some extent, professionals during these early stages focused on wider educational and social needs and can therefore be seen to reflect the aims of the BSF agenda.

Although during Stage 4 there were fewer apparent tensions, we have argued that this was because the motive of many of the professional groupings at this stage was to win the bid. This led to concerns from some over the competitive process and their ability to give proper guidance in this context.

In Stage 5, the delivery team was selected and the agreed design had to be rigorously tested for affordability. Consultation with the end-users was described by both parties as disappointing; the design was fixed at this point and active collaboration could not take place.

The contractors were under commercial pressures to deliver the school within a tight programme. Key decisions were made at this stage to cut costs that impacted on the quality and performance of the built design. Hence the aims of the educational vision were compromised:

> The idea, the enthusiasm, the desire by people to make something new and exciting and make it work really well was a major driver of the design. As a consultant employed by the contractor, it was very difficult to influence this process, and the only way that I can see that alternative views could be put forward would be from within the client organisation, within the group of people who ultimately decided what kind of school they wanted.
>
> *(Acoustic consultant)*

The council's dominant motive for innovation and educational transformation propelled the process forward, even though government guidance (BB93 and BB98) showed concern over the ability of open-plan

TABLE 4.2 Summary of main points for each stage of the vision/design/build process

	Stage	Dominant Motives	Tensions/Trade-Offs (which decisions were abandoned and which carried forward)
1	Introduction of BSF – school building agenda	DfES: drive social and educational reform	None relating to the design process
2	Local recontextualisation	Council – improve attainment; transform teaching & learning school – improve attainment/environment	Some resistance from school to continue traditional ways of teaching and learning; school adopted council's ideas on personalisation
3	Transforming the educational vision into a design	Council/School (motives as above) project management company – deliver project on time/within budget/to agreed output concept architects: maintain practice reputation as good designers	Concept architects adopted STT's ideas on personalisation Collaborative approach to multi-agency working developed and appeared to be working.
4	Finding the design and build team – invitation to tender; invitation to continue dialogue; invitation to submit final bid	Council/School (to find architect to deliver vision) Partnership for Schools: deliver project on time/in budget Consortium leader: to win bid Contractors: to win bid Design consultants: to win bid Education advisors: to win bid ICT provider: to win bid FM Services: to win bid	Preferred design not chosen due to other commercial considerations Tension in motives: financial versus vision Tension with communication: design consultants were not able to communicate directly with the end users – collaborative multi-agency working Tension – concept designs developed in Stage 3 between council/school/concept architects represented the vision developed in Stage 2 but bidding architects chose to start new designs LEP formed
5	Design and build	Council/School: vision translated into material design PfS – little evidence of involvement Consortium leader: deliver project on time and in budget to agreed output Contractors: as consortium leader plus a commercial interest Design consultants: to maintain reputation and produce an innovative design ICT services FM services	LEP formed Tension – reduced input from school and council Tension – delivery architects prevented from orchestrating collaborative working due to procurement process Budget and time constraints resulted in reduced specifications
6	Process of occupation		
7	Space to place		

Note: Design consultants refers to: delivery architects; acoustic engineers; M and E consultants; structural engineers; landscape designers.

Credit: Feilden Clegg Bradley Studios

spaces to deliver effective environments for teaching without enhanced specifications. The council failed to follow guidance and implement the aims of the original educational vision effectively.

From our findings of the visioning process, the evolution of the educational vision and how and by whom it was developed, through the stages of a school building process, impacted greatly on the effectiveness of the teaching and learning environment for its end-users. The complex organisational and technological processes of school building required constant assessment of how the implementation of the educational vision was carried out and how key decisions influenced the quality of the built environment and its flexibility and adaptability to support changing pedagogies and community needs through the lifecycle of a school building.

The council's commitment to their educational vision driven by BSF's drive for transformation often overshadowed technical guidance by DfES and design consultants to critically assess how innovative designs can effectively support new ways of learning and be able to adapt to future changes in educational policies and teaching practices.

In addition, following BSF's guidance on wide consultation, the council positively engaged the school and its wider community at the preliminary visioning stage but disengaged them from the process at the implementation stage. Furthermore, routes of communication during this stage between the design consultants and the end-users were actively discouraged by the delivery team, which disempowered both parties from the potential to work collaboratively towards a shared goal.

While it is common practice in the construction industry to commission architects to develop initial concepts and design briefs, and then subsequently commission a different design team to deliver the project, it is appropriate to question, in times of austerity, whether this is the most efficient use of available resources, given the problematic outcomes detailed here. Maintaining continuity between different design stages and different stakeholders and preserving leadership is essential to delivering the educational vision of the project throughout the design and construction of a school.

Conclusion

We have analysed the output of the 'visioning process' at the inception of the project, the clear articulation of that vision to the delivery team, and how that vision is translated into material form. The rationale for conducting this phase of our research was the argument that there is a limited understanding of how educational visions are developed and translated into material designs. The reason we felt it important to address this is that in order to understand teachers', students' and parents' perceptions of school space it is necessary to understand what that space was intended for and how it was intended to be used.

We found that while plenty of time was allocated to the development of an educational vision in the case of School C4 and while all agencies involved in the design process during Stages 1–3 worked relatively collaboratively, motives did differ – although not to the extent that these would cause a discontinuity. Problems arose when it came to translating the educational vision into a material design due to the disconnect between groups of professionals in Stages 4 and 5 and due to the fact that there did not appear to be anyone driving the vision through the latter stages.

A further problem related to the lack of expertise of the council and STT in relation to whether their educational vision could be translated into an effective material design – in other words, would it work in the way intended? The overwhelming drive for personalisation and open-plan spaces meant that advice given in BB93 and by the acoustics engineers, for example, that the specification of open-plan learning spaces would not comply with the acoustic performance criteria, fell on deaf ears.

The BSF process has been criticised by many for its overly complex procurement process and funding structure (NAO, 2009). From our detailed analysis of School C4's design process, we would argue that the extent to which an educational vision can be translated into a material design does not hinge on the efficiencies of the procurement process per se but on the development and implementation of the educational vision. In order to achieve this we argue for the need to adopt lessons from Daniels et al. (2007), which, while not focusing specifically on school design, does share many similarities in relation to problems experienced in multi-agency working.

In both cases the architectural and teaching professions fail to prepare practitioners for working outside established organisational practices. This is particularly problematic when the boundaries between different organisations are well established and strong. In order to blur these boundaries, Daniels et al. (2007) would suggest that practitioners need to:

- Be receptive to how other types of practitioner are interpreting the trajectory of the an educational vision and be receptive to the expertise which informs those interpretations;
- Engage with local systems of distributed professional expertise;
- Follow the design and procurement trajectory in a fluid and responsive way outside of their own established organisational systems;
- Analyse the suitability of organisational conditions for developing this form of work.

(Daniels et. al., 2007, p. 1)

A second research aim was to develop a methodology for exploring the process of school design in relation to educational vision. The analysis conducted allowed us to explore the process of design across different periods. We found that motives can change depending on the aims and objectives at a particular point in time; when the motives of different professional groupings differ at particular stages this can cause tensions. In order for the design and build to proceed, trade-offs had to occur. Early indications, based on observation and interview data, are that because of these trade-offs spaces have been built which do not align with the original goals established during Stages 2 and 3.

This analysis has provided us with a stage model for analysing the process of developing an educational vision and translating this into material spaces. The next step in this project is to apply this stage framework to the other four schools involved in our project. We will explore similarities and differences in the process of designing these schools, some of which were procured under BSF and some as Sponsored Academies/PFI programmes, with the aim of developing a model for understanding the process of school design and the relationship of this process to the educational experience of end-users. This will allow us to make recommendations for effective multi-agency work between all the professional organisations involved in school design (e.g. educationalists, architects, contractors, CABE) and thus enhance the practice of designing, engineering and facilitating learning spaces for changing pedagogical practices to support a mass education system, and greater student diversity.

References

Ariani, M.G. and Mirdad, F. (2016) The effect of school design on student performance, *International Education Studies*, 9(1): 175–81.

Audit Commission (2003) *PFI in Schools*. London: Audit Commission.

Baars, S., den Brok, P., Krishnamurthy, S., Joore, J.P. and van Wesemael, P.J.V. (2018) Constructing a framework for the exploration of the relationship between the psychosocial and the physical learning environment. In *Transitions Australasia: What is needed to help teachers better utilize space as one of their pedagogic tools* (pp. 90–97).

Barrett, P., Davies, F., Zhang, Y. and Barrett, L. (2017) The holistic impact of classroom spaces on learning in specific subjects, *Environment and Behavior*, 49(4): 425–51.

Barrett, P., Zhang, Y., Moffat, J. and Kobbacy, K. (2013) A holistic, multi-level analysis identifying the impact of classroom design on pupils learning, *Building and Environment*, 59: 678–89.

Bazerman, C. (1997) Discursively structured activities, *Mind, Culture and Activity*, 4(4): 296–308.
Brittin, J., Frerichs, L., Sirard, J., Wells, N., Myers, B., Garcia, J., Sorensen, D., Trowbridge, M. and Huang, T. (2017). Impacts of active school design on school-time sedentary behavior and physical activity: A pilot natural experiment, *PLOS ONE*, 12(12).
Burke, C. and Grosvenor, I. (2003) *The School I'd Like. Children and Young People's Reflections on an Education for the 21st Century*. London: RoutledgeFalmer
Burke, C. (2010) About looking: Vision, transformation, and the education of the eye in discourses of school renewal past and present, *British Educational Research Journal*, 36(1): 65–82.
Burke, C. and Grosvenor, I. (2008) *School*. London: Reaktion.
CABE (2004) Being involved in school design: A guide for school communities, local authorities, funders and design and construction teams. http://webarchive.nationalarchives.gov.uk/20110118095356/http:/www.cabe.org.uk/files/being-involved-in-school-design.pdf
CABE (2007) Creating excellent secondary schools: a guide for clients. http://webarchive.nationalarchives.gov.uk/20110118095356/http:/www.cabe.org.uk/files/creating-excellent-secondary-schools.pdf
Can, E. and İnalhan, G. (2017) Having a voice, having a choice: Children's participation in educational space design, *The Design Journal*, 20(sup1): S3238 – S3251.
Cartwright, N. (2007) Evidence-based policy: So, what's evidence? Enquiry, evidence and facts: An interdisciplinary conference. www.britac.ac.uk/events/2007/evidence/abstracts/cartwright-long.cfm
Chidgey, J. (2009) Building schools for the future: Implications for design and technology. www2.futurelab.org.uk/resources/publications-reports-articles/web-articles/Web-Article1316
Cleveland, B. and Fisher, K. (2014) The evaluation of physical learning environments: A critical review of the literature, *Learning Environments Research*, 17(1): 1–28.
Cleveland, B. and P. Soccio (forthcoming), *Research Report: Development and Pilot Testing of the School Spaces Evaluation Instrument (SSEI): Module 2 – Technical Performance/Indoor Environment Quality (IEQ) Module Alignment of Pedagogy and Learning Environments*. Melbourne: University of Melbourne.
Daniels, H., Leadbetter, J. and Warmington P. with Edwards, A., Brown, S., Middleton, D., Popova, A., and Apostolov, A. (2007) Learning in and for multiagency working, *Oxford Review of Education*, 33(4): 521–38.
Den Besten, O., Horton, J. and Kraftl, P. (2008) Pupil Involvement in School (Re)Design: Participation in Policy and Practice, *CoDesign*, 4(4): 197–210.
de Vrieze, R. and Moll, H. (2016) An analytical perspective on primary school design as architectural synthesis towards the development of needs-centred guidelines, *Intelligent Buildings International*, 1: 1–23.
DfE (2013) Priority School Building Programme. www.education.gov.uk/schools/adminandfinance/schoolscapital/priority-school-building-programme
DfES (2003) *Building Schools for the Future: Consultation on a New Approach to Capital Investment*. London: DfES.
DfES (2004) *Building Schools for the Future: A New Approach to Capital Investment*. London: DfES.
Edwards, A., Daniels, H., Gallagher, T., Leadbetter, J. and Warmington, P. (2008) *Improving Interprofessional Collaborations in Children's Services*. London: Routledge.
Elledge, J. (2012) And the medal for worst cock-up of all goes to . . ., *The Guardian*, 14 August.
Engeström, Y. (1999) Innovative learning in work teams: Analysing cycles of knowledge creation in practice, in Y. Engestrom, R. Miettinen and R.L. Punamäki (ed.), *Perspectives on Activity Theory* (pp. 377–406). Cambridge: Cambridge University Press.
Engeström, Y., Brown, K., Christopher, C. and Gregory, J. (1997) Coordination, cooperation and communication in courts: Expansive transitions in legal work, *The Quarterly Newsletter of the Laboratory of Comparative Human Cognition*, 13(4): 88–97.
Engeström, Y., Engeström, R. and Vähäaho, T. (1999). When the center does not hold: The importance of knotworking, in S. Chaiklin, M. Hedegaard and U.J. Jensen (eds.), *Activity Theory and Social Practice: Cultural-Historical Approaches* (pp. 345–74). Aarhus, Denmark: Aarhus University Press.
Foxell, S. and Cooper, I. (2015). Closing the policy gaps. *Building Research & Information*, 43(4): 399–406. Retrieved from http://dx.doi.org/10.1080/09613218.2015.1041298
Frerichs, L., Brittin, J., Intolubbe-Chmil, L., Trowbridge, M., Sorensen, D. and Huang, T. (2015) The role of school design in shaping healthy eating-related attitudes, practices, and behaviors among school staff, *Journal of School Health*, 86(1): 11–22.
Galasiu, A.D. and Veitch, J.A. (2006) Occupant preferences and satisfaction with the luminous environment and control systems in day lit offices: A literature review, *Energy and Buildings*, 38: 728–42.

Gilavand, A. and Hosseinpour, M. (2016). Investigating the impact of educational spaces painted on learning and educational achievement of elementary students in Ahvaz, southwest of Iran, *International Journal of Pediatrics*, 4(2): 1387–96.

Goodenow, C. (1993). The psychological sense of school membership among adolescents: Scale development and educational correlates, *Psychology in the Schools*, 30: 79–90.

Greene, J.C. (2008) Is mixed methods social inquiry a distinctive methodology? *Journal of Mixed Methods Research*, 2(1): 7–22.

House of Commons: Education and Skills Select Committee (2007) *Sustainable schools: Are we building schools for the future? Seventh Report of Session 2006–07*. London: Stationery Office.

House of Commons Public Accounts Committee (2009) *Building Schools for the Future: renewing the secondary school estate*.

Hygge, S. (2003) Classroom experiments on the effects of different noise sources and sound levels on long-term recall and recognition in children, *Applied Cognitive Psychology*, 17: 895–914.

Imms, W. and Byers, T. (2016) Impact of classroom design on teacher pedagogy and student engagement and performance in mathematics, *Learning Environments Research*, 20(1): 139–52.

ITCD Bid Document (2008) Vol 1 D(a) Part 1 Design.

James, S. (2011) *Review of Education Capital*. London: DfE.

KCC (2005) *Template 01 – Schools for the Future*. Kent County Council and Office of the Deputy Prime Minister.

Kerosuo, H. (2015. BIM-based collaboration across organizational and disciplinary boundaries through knotworking. *Procedia Economics and Finance*, 21: 201–8.

Könings, K., Bovill, C. and Woolner, P. (2017) Towards an interdisciplinary model of practice for participatory building design in education, *European Journal of Education*, 52(3): 306–17.

Kraftl, P. (2011) Utopian promise or burdensome responsibility? A critical analysis of the UK government's Building Schools for the Future Policy, *Antipode*, 44(3): 847–70.

Lau, J., Wang, L.M., Waters, C. and Bovaird, J. (2016) A need for evidence-based and multidisciplinary research to study the effects of the interaction of school environmental conditions on student achievement, *Indoor and Built Environment*, 25(6): 869–71.

Leont'ev, A.N. (1978) *Activity, Consciousness and Personality*. Englewood Cliffs: Prentice Hall.

Livesey, P. (2012) *Priority School Building Programme*. London: DfE.

MacPhail, A. (2001) Nominal group technique: A useful method for working with young people. *British Educational Research Journal*, 27: 161–70.

Magzamen, S., Mayer, A.P., Barr, S., Bohren, L., Dunbar, B., Manning, D., . . . and Cross, J. E. (2017) A multidisciplinary research framework on green schools: Infrastructure, social environment, occupant health, and performance, *Journal of School Health*, 87(5): 376–87.

Mahony, P., Hextall, I. and Richardson, M. (2011) 'Building Schools for the Future': Reflections on a new social architecture, *Journal of Education Policy*, 26(3): 341–60.

Mahony, P. and Hextall, I. (2013) Building Schools for the Future: 'Transformation' for social justice or expensive blunder? *British Educational Research Journal*, 39(5): 853–71.

Mäkelä, T. (forthcoming, 2018) Developing a conceptual framework for participatory design of psychosocial and physical learning environments.

Mäkelä, T. and Helfenstein, S. (2016) Developing a conceptual framework for participatory design of psychosocial and physical learning environments, *Learning Environments Research*, 19(3), 411–40.

Mandelson, P. (1997) *Labour's next steps: Tackling social exclusion*. Fabian pamphlet no. 581. London: The Fabian Society.

Maxwell, L.E. (2016). School building condition, social climate, student attendance and academic achievement: A mediation model, *Journal of Environmental Psychology*, 46, 206–16.

Ministry of Education (2014) Modern learning environments. www.minedu.govt.nz/NZEducation/Education Policies/Schools/PropertyToolBox/StateSchools/Design/ModernLearningEnvironment/M LEDQLSStandards.aspx

Nambiar, R.M.K., Nor, N.M., Ismail, K. and Adam, S. (2017) New learning spaces and transformations in teacher pedagogy and student learning behavior in the language learning classroom, *The Southeast Asian Journal of English Language Studies*, 23(4): 29–40.

National Audit Office (2009) *The Building Schools for the Future Programme: Renewing the Secondary School Estate*. Report HC 135 Session 2008–2009.

Newman, D., Griffin, P. and Cole, M. (1989) *The Construction Zone: Working for Cognitive Change in School*. Cambridge: Cambridge University Press.

OECD (2009) *International Pilot Study on the Evaluation of Quality in Educational Spaces (EQES) User Manual Final Version.* Paris: OECD.
OECD (2013) *Innovative Learning Environments.* Paris: OECD.
OECD (2014) *Effectiveness, Efficiency and Sufficiency: An OECD Framework for a Physical Learning Environments Module.* Paris: OECD.
Ofsted (1999–2000) Annual report. www.ofsted.gov.uk/resources/annual-report-19992000-ofsted-subject-reports-primary
Pantidi N. (2016) Supporting fluid transitions in innovative learning spaces: Architectural, social and technological factors, in *Architecture and Interaction.* London: Springer.
Pearson, R. and Howe, J. (2017) Pupil participation and playground design: Listening and responding to children's views, *Educational Psychology in Practice*, 33(4): 356–70.
PfS/4ps (2008) *An Introduction to Building Schools for the Future.* London: 4ps and PfS
RIBA (2007–08) Outline plan of work 2007. www.architecture.com/Files/RIBAProfessionalServices/Client Services/RIBAOutlinePlanOfWork2008Amend.pdf
RIBA (2013) Outline plan of work 2013. www.architecture.com/TheRIBA/AboutUs/Professionalsupport/RIBAOutlinePlanofWork2013.aspx#.Uic3AqBOSJo
Sailer, K. and Penn, A. (2010). Towards an architectural theory of space and organisations: Cognitive, affective and conative relations in workplaces. *2nd Workshop on Architecture and Social Architecture*, EIASM, Brussels, May. http://discovery.ucl.ac.uk/1342930.
Schools for the Future (2004) Exemplar Designs Compendium. https://webarchive.nationalarchives.gov.uk/20110113135138/www.teachernet.gov.uk/_doc/6113/Exemplar%20Designs%20compendium.pdf.
Shaoul, J., Stafford, A. and Stapleton, P. (2010) Financial black holes: The disclosure and transparency of privately financed roads in the UK, *Accounting, Auditing and Accountability Journal*, 23(2): 229–55.
Shaughnessy, R. J., Haverinen-Shaughnessy, U., Nevalainen, A. and Moschandreas, D. (2006) A preliminary study on the association between ventilation rates in classrooms and student performance, *Indoor Air*, 16(6): 465–8.
Sigurðardóttir, A.K. and Hjartarson, T. (2016) The idea and reality of an innovative school: From inventive design to established practice in a new school building, *Improving Schools*, 19(1), 62–79
Strauss, A. and Corbin, J. (1998) *Basics of Qualitative Research*, 2nd edn. Newbury Park, CA: Sage.
UNISON (2003) *What is Wrong with PFI in Schools?* A PFI report for UNISON.
Victor, B. and Boynton, A.C. (1998). *Invented Here: Maximizing Your Organization's Internal Growth and Profitability.* Boston, MA: Harvard Business School Press.
Vygotsky, L.S. (1987) *The Collected Works of L.S. Vygotsky. Vol. 1: Problems of General Psychology, Including the Volume Thinking and Speech*, ed. R.W. Rieber and A.S. Carton, trans. N. Minick. New York: Plenum Press.
Whyte, B., House, N.H. and Keys, N. (2016) Coming out of the closet: From single-cell classrooms to innovative learning environments, *Teachers and Curriculum*, 16(1), 81–8.
Wilson, H. K. and Cotgrave, A. (2016) Factors that influence students' satisfaction with their physical learning environments, *Structural Survey*, 34(3): 256–75.
Winterbottom, M. and Wilkins, A. (2009) Lighting and discomfort in the classroom, *Journal of Environmental Psychology*, 29(1): 63–75.
Woolner, P., Hall, E., Higgins, S., McCaughey, C. and Wall, K. (2007) A sound foundation? What we know about the impact of environments on learning and the implications for Building Schools for the Future, *Oxford Review of Education*, 33(1): 47–70.

Credit: HKS Architects

5
DESIGN AND PRACTICE

In Chapter 4, we described how we developed a model for exploring the complex processes of school design and occupation. Architecture is a deeply social practice, but in contemporary architecture, the social dimension of design and analysis is often absent. Architecture students learn how to render and sculpt spacial forms, but less time is spent on considering the far less intangible component of human interaction and the social structures that design can support. The most explicit view of how some contemporary architects perceive architectural spaces can be glimpsed in countless architectural journals that overwhelmingly display abstract, sculptural minimalist interiors bereft of human bodies and their living traces. While some architects may consider 'flow' or other abstract concepts that examine how inhabitants move and navigate through spaces; for example, the development of Space Syntax by Hillier et al. (1989; Hillier, 2007) have attempted to quantify the spatial and configurational logic of buildings and cities in order to understand the influence of spacial configuration on social life. Ratti (2004) have pointed to the limitations of space syntax tools in the analysis of certain geometrical configurations. In times of austerity for school design, efficiency and standardisation is often prioritised over social concerns in the local context.

Yet according to our research on learning spaces, the most important component that influences the experiences of students and teachers is social relations (discussed further in Chapter 6). Everyone from the designers themselves to policymakers who commission the buildings and those who manage the development of the projects should therefore foreground the social in school design from the very beginning. Everyday problems such as anxiety, bullying, poor behaviour absenteeism, staff retention and atrophied student engagement that arise in schools spring from pressures and frictions in communities and vulnerabilities in social relations. If school design is viewed as an integrated component of the holistic school environment, then perhaps social problems can be influenced and mitigated by socially engaged school design. A school is not simply a shell of designated dimensions containing students and teachers but an important design opportunity to support the social structure of school communities in one of the most vital public spaces for youth and those who work with them.

To this end, in this chapter we will describe the key characteristics of the vision, design, construction and occupation of 10 English secondary schools in five localities using the pedagogic post-occupancy analysis of the stage model developed in Chapter 4. This material on the social dimensions of school design will provide a holistic understanding of how design processes impact on end-users' experiences of schooling.

Locality A: transformation through community engagement

Tony Blair opened England's first city academy, the Business Academy Bexley, in 2002, it was heralded as one of the symbols of New Labour's education policy. The first wave of academies was built in deprived inner-city areas, part financed by the private sector and were intended to transform failing schools, taking them out of local authority control. These social and educational aims were supported by a radical rethink of traditional school design, creating buildings that are open to the community and highlight the academy's specialist subject areas. Bexley's design applied the philosophy of the Foster and Partners' office buildings, which pioneered the humanisation of the workplace in the early 1970s, to the design of a business academy that would engender a sense of community and institutional pride. The curriculum places particular emphasis on business, art and technology, driven in part by Bexley's entrepreneurial sponsors. Within the building the different subject areas are the focus of three light-filled internal courtyard spaces. Three levels of open classrooms are arranged around the courtyards, maintaining a visual connection across the thematic zones and allowing natural supervision. On the upper levels, some of the spaces are open on one side to create flexible, screened teaching areas within the larger volume. The business theme continues with a miniature stock exchange, complete with plasma screens, which gives students a taste of trading, strengthening Bexley's connection with the towers of Canary Wharf. This was a bold new design for new ways of delivering education.

In parallel with this development, as soon as the Academy Programme was announced by the Labour government in 2000, the headteacher(1) of the predecessor school in Locality A realised that this was an opportunity to change the educational prospects for students in his city. The predecessor school and the new city academy serve local communities, many of which are socially disadvantaged. Over two-thirds of students are from minority ethnic groups, with over 40 per cent learning English as an additional language. The challenges of setting up a city academy were enormous. First the school had to find sponsors, and after writing hundreds of letters the two sponsors, the University of the West of England and the city's football club came on board. Then the headteacher(1) held 148 public consultations with politicians, local communities and parents. The headteacher(1) was always in the 'driving seat' and knew from the outset that he wanted to build a new school in which young students could feel a real sense of belonging, if they were to learn effectively. He wanted to create a school which felt small even though the number of students would be around 1,300. The design vision centred around five learning 'villages', breaking down the academy into cohesive communities, promoting a strong sense of belonging and responsibility among students, teachers and parents. Throughout the process the headteacher(1) worked closely with the architects, and the design became a blueprint for the building of new secondary schools across England:

> The design concept is something that came through very clearly from the headteacher. It was a big school, when it opened, there were 1,236 students but clearly that's a large group of people and within that group you don't get any sense of belonging and protection and you could get lost in the crowd. The idea of breaking down the school into a group of units, almost like schools within schools was really important from the beginning for the principal.
>
> *(Architect interview)*

> It's all linked together as one building, five different buildings with four learning villages towards the back of the site and then the community village at the front of the street frontage. And we called them villages [Figure 5.2] – again that's something that came from the headteacher – because the

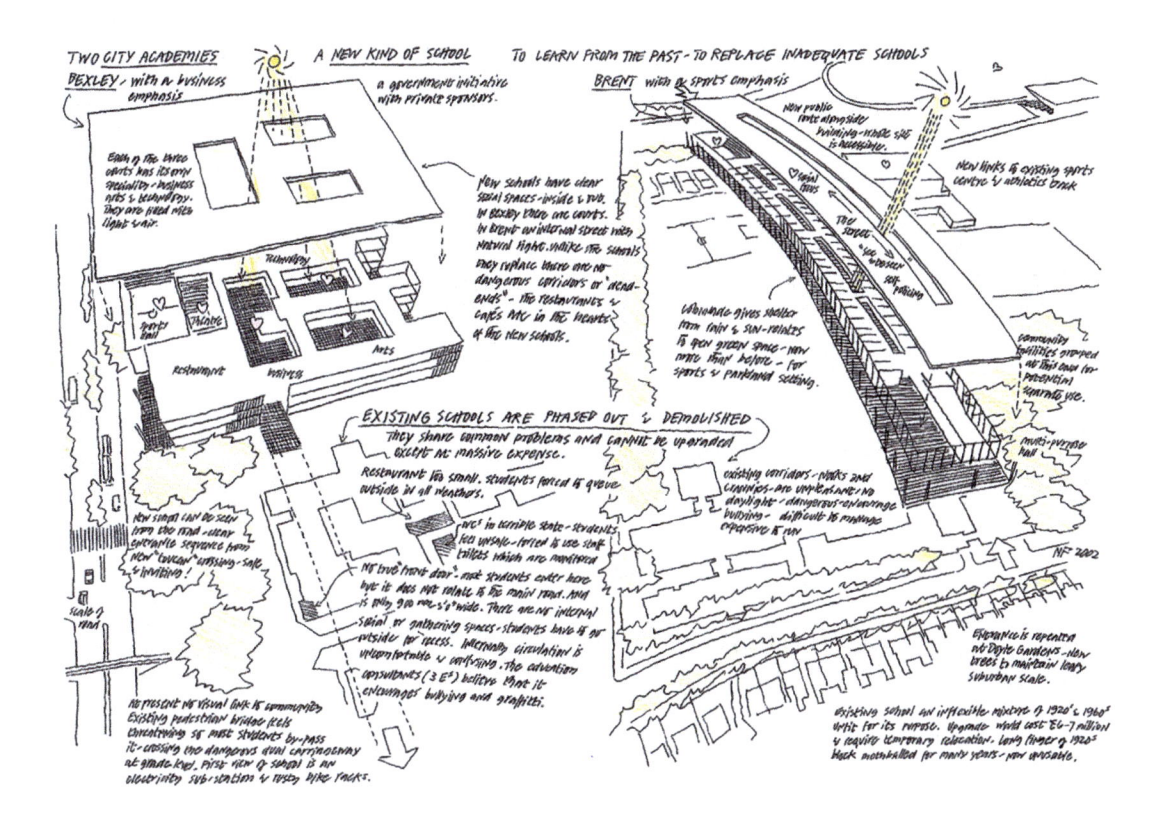

FIGURE 5.1 Bexley Academy
Credit: Fosters and Partners

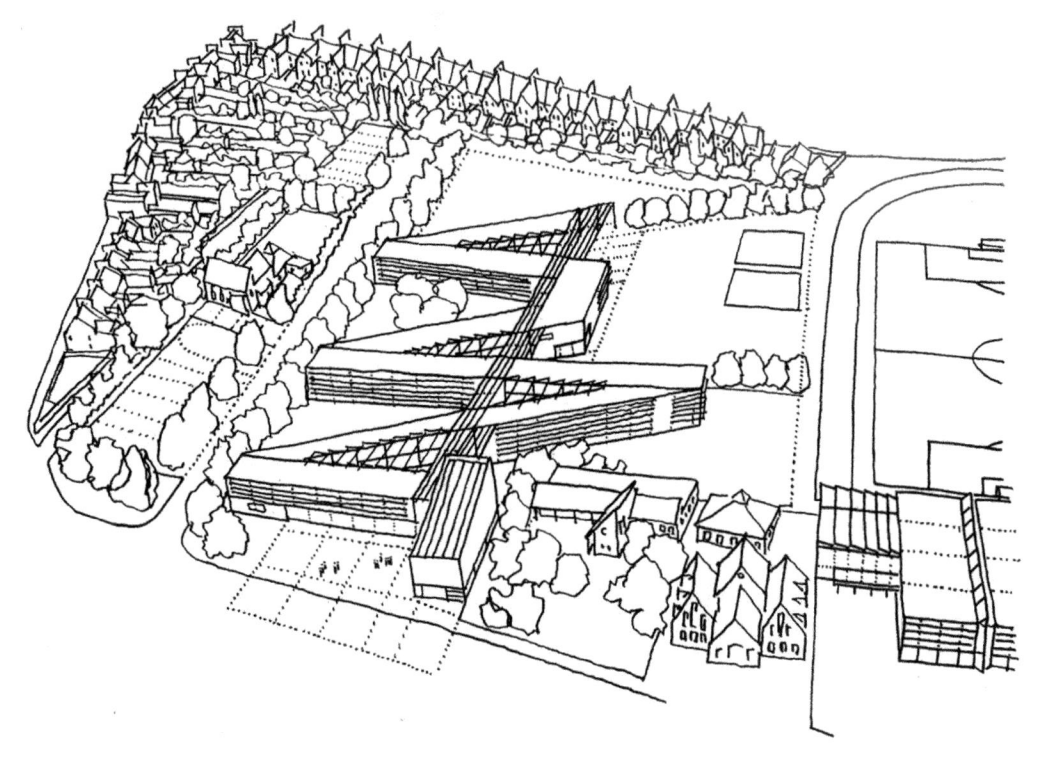

FIGURE 5.2 Perspective view of initial design concept
Credit: Feilden Clegg Bradley Studios

idea was that they'd be a kind of a pastoral base for all of the students who are here, so they had this space where they felt they belonged as opposed to being part of the wider group.

(Architect interview)

This is a typical learning village [Figure 5.2], each one contains social hub space which is a breakout space for the activities of the classrooms. There are drawings of the school from those early concept drawings where we started to pull apart the standard corridor with teaching rooms on both sides, and that turned into the V-shaped clusters. It's how we get light down into the ground floor and avoid the dark corridors that the principal reacted so incredibly strongly against in other schools we visited.

(Architect interview)

I feel that the way we worked on this building is that it's kind of the crystallisation of the social structure on the site and the building is representative of that, which while Ray didn't focus on the architecture, he didn't tell us what it should look like that, he knew how the building should work.

(Architect interview)

This city academy was the first academy built in south-west England and became much more than a secondary school. Under the headteacher's leadership it offered joint service provision with teenage health services and social care, alongside full representation at neighbourhood management boards. It also pioneered the

FIGURE 5.3 Perspective view of typical village cluster
Credit: Feilden Clegg Bradley Studios

opening of a full employment service, management of a community centre, and provision of a day centre for adults with learning difficulties. The new city academy became a focal point for community cohesion and gave the community a real sense of belief and confidence in their local school. Under the headteacher's leadership the number of students following on to higher education increased significantly within five years of the academy opening. The 2009 Ofsted report of city academy said: 'The principal's leadership is outstanding. He has been instrumental in raising the aspirations of students and in placing the Academy at the heart of the local community. Through its strong partnerships with other agencies, the Academy has placed a strong accent on promoting cohesion by encouraging students to value participation in local life, understand and appreciate diversity as well as to have high self-esteem and expectations of themselves as learners.'

> This school wouldn't have looked like this had the Principal not been involved, he was an integral member of the design team and leading the ideas. I guess the obvious advantage of that is you end up with a school that is very tailored to his particular requirements of what he wants the school to be. And the downside of course is that he might not be around . . . I guess he was around for about five years and then it was handed on, so how does it work afterwards, you kind of almost need people who come in subsequently to buy in to those original concepts or you've created something that's a strait jacket for the way the school might develop later on. Having said that, I think the ideas that came forward work successfully at quite a fundamental level
>
> *(Architect's interview)*

Continuity in the design process was cemented in the structure of the procurement method. The traditional two-stage tender was led by the design team, the architect was the lead consultant and responsible for co-ordination and quality control of the project from inception to completion. All the design consultants were appointed directly by the client – the school – and contractually obligated to meet the needs of the end-users. The design team had a close working relationship with the school led by the headteacher with a strong educational vision. The procurement and design process allowed the head to clearly articulate his educational vision to the design team and critically he was fully involved in the progression of the design and the key decisions that may impact on the intent of the proposed educational vision. The close collaboration between the school and the design team also ensured that the headteacher could communicate his ideas directly to each design team member and continuously respond to issues or concerns:

> The headteacher is paying and spending £20,000 a year on paint, and that's not painting, it's just paint to keep the place looking pristine. That was part of the idea from the very beginning, and as soon as the building opened there was a member of staff who was tasked with going round every day and painting over the worst of the scuff marks, then during the weekend there would be more of a comprehensive repaint. And clearly that's still part of the ethos now.
>
> *(Architect interview)*

The headteacher(1) left five years after occupation of the new school building. Headteacher(2) tried to follow the pedagogic vision of the first occupation but the increase in incidents of poor behaviour and vandalism meant a progressively stricter control of students' movements and use of social spaces in the school. Social Hub spaces were no longer being used for individual or group independent study or during break times. IT and computer facilities were taken away due to vandalism and antisocial student behaviour. In 2013, the Department of Education issued a pre-notice warning due to unacceptably low academic performance and headteacher(2) left. Headteacher(3) struggled to raise academic standards, and in 2015 the school was placed into special measures status.

In Locality A, the design, construction and occupation processes were characterised by high levels of collaboration led by the headteacher with a clear educational vision. This was a form of mutual shaping of the emergent design by the school and the design and construction team. But over time, the second and subsequent occupations were unable to use the design as envisioned, the design disjointed from the educational practice as enacted.

In Locality C, we describe a case study of five schools (Figure 5.4) designed for one local authority for Wave 3 of the Building Schools for the Future Programme which provides a clear account of how one design can be used and experienced in different ways.

FIGURE 5.4 Locality C, Schools C1–C4 and Comparator School CC1

Locality C

Personalisation and collaboration

In 2004, David Milliband gave a speech titled 'Choice and Voice in Personalised Learning' as the UK Schools Minister at the OECD where he outlined five components of personalised learning to guide policy development:

1. It needs assessment for learning and the use of data and dialogue to identify every student's learning needs.
2. It calls for the development of the competence and confidence of each learner through teaching and learning strategies which build on individual needs.
3. It requires curriculum choice which engages and respects students, while allowing for breadth of study, personal relevance, and clear paths through the system.
4. It demands an approach to school organisation and class organisation based around student progress. Workforce reform is a key factor, and the professionalism of teachers is best developed when they have a range of adults working with them to meet diverse student needs.
5. Personalised learning means the community, local institutions and social services supporting schools to drive forward progress in the classroom.

To deliver Personalised Learning required new forms of school designs. Miliband in February 2003 announced that 'School buildings should inspire learning. They should nurture every pupil and member of staff. They should be a source of pride and a practical resource for the community.' In the 2003 DfES consultation document, the Labour government suggested that 'Every child deserves to learn in modern school buildings with state-of-the-art facilities. We want local communities to ensure that over the next 10 to 15 years all secondary pupils can learn in modern accommodation, fully suited to their needs and to the challenges of the 21st Century' (DfES, 2003), which led to the start of the UK's Building Schools of the Future Programme in 2004.

National policy to local recontextualisation

In parallel, two years before BSF was announced, the county council initiated an internal review of secondary provision and, in partnership with the schools, developed a strategy to reshape the county's educational landscape in line with BSF's 'Transformation' Agenda (DfES, 2003). In order to develop the council's secondary strategy, and ensure that transformation occurred, they recruited a Secondary Transformation Team (STT) of ex-headteachers and consultant educationalists. The National Audit Office's review of the BSF programme in 2009 commended this county's innovative and cohesive educational strategy prior to entry into the BSF programme (NAO, 2009). This was an initiative in which it was argued that a novel artefact – a school building – could function as a tool of transformation projected forward in time from the point at which it was envisioned.

From the onset, there was a clear stated national policy priority to use BSF to deliver educational transformation but there was no evidence of a 'coherent definition' of this central objective (James, 2011, p. 12). The lack of guidance led to variations in the programme, process, design and outcome as each of the bodies involved had to implement their own interpretation of 'educational transformation' (NAO, 2009, p. 6; James, 2011, p. 13). Local recontextualisation of national policy took place.

The STT set out to encourage the county's schools to work collaboratively in order to improve standards through the transformation of local practices. This work involved the development for the Schools

Improvement Partner Programme for the council in 2006. This was a local statement of the meaning of the national BSF initiative:

> We had to understand what transformation meant to us?
>
> *(STT interview)*

This process led to the articulation of a local educational vision. The STT recognised they needed to address priorities in the local contexts. They involved headteachers from each of the eight schools involved in the county's BSF programme (Wave 3 and Wave 4) in developing their own educational vision. Numerous workshops were held with members of the schools, the STT and the council.

Four areas were identified as core elements for the county's BSF programme:

1. Transforming the organisation of learning in schools.
2. Developing capacity and structures by extending collaborative partnerships between schools by developing clusters and transforming them into Education Improvement Partnerships.
3. Placing schools firmly at the heart of their local community.
4. Designing and developing the learning infrastructure in collaboration with the sample schools.

The council developed their new education vision with the publication of the council's Template for Schools of the Future document in 2005. The vision to deliver personalisation demanded a new educational space model. The concept aimed to teach students in 'learning clusters of schools within school' to break down the scale of the whole student population, in order that students can be nurtured in smaller groups. The new space model demanded the flexibility to adapt to different learning approaches from university style lectures for 240 students to small group discussions and independent IT led learning modules for one. The council's rationale for change was grounded in an intention to replace traditional didactic teaching methods in which students were assumed to be passive recipients of knowledge to one in which students became actively engaged with the learning experience through inquiry-based learning. The council's strategy for transformation was to provide the vision for new ways of delivering secondary education in the county. It followed the core elements of the BSF programme, which promoted personalised learning in curriculum content, assessment, learning style and different forms of learning. Models were developed for a diverse range of open-plan learning spaces which illustrated features of adaptable environments, addressing acoustic and ICT needs for learning in a variety of group sizes, from individual learning to large group sessions. These design templates would serve as models for how personalised learning could be delivered in the council's new BSF schools.

It was apparent that the STT and council was strongly committed to this new educational vision and sought to ensure that it was championed by each school and their design teams. Each school, led by the headteacher, worked with the sample schools design team and the council to develop their own learning clusters of specialist areas within the council's concept set out in their template. They argued for designs that could not be used for traditional forms of practice. The argument was that design would change practice:

> These buildings have been designed in a way that makes it almost impossible for them to go back to square one. That was our guiding principle.
>
> *(STT interview)*

From March 2006, the school communities worked with the council and design teams to transform the new educational vision into school designs that could be developed into exemplar design models for the replication all over the county. The BSF procurement process was overly complex and strongly echoed the findings of the James report (James, 2011). At financial close, both the bidder and the authority will have reached agreement on all the contractual documents, in addition to all relevant technical issues, in addition to all other matters affecting the unitary charge and the interest rate incurred on the bank debt taken out to finance the project. The legal agreements will be signed by all parties and then held until financial close.

The bidders reported that this onerous process placed 'unsustainable strain' on their bidding capacity (NAO, 2009, p. 47). Here we obtained evidence of the early stages of tension and strain on the knotworking and co-configuration processes.

The council and the schools had to invest intensive periods of time in the competitive dialogue process and the participants we interviewed felt they were being asked the same questions they had answered in earlier phases of the project. The outcome of the process was also 'disappointing' for some of the schools, as the consortium that produced their preferred designs did not win the bid. Interview and documentary data suggest that different object/motives were at play.

The delivery architects for the preferred bidder informed us of their 'frustrating' experience of discontinuities and contradictions in the design process during this phase. This was as a result of conflicts between their professional role to understand and design for the needs of the end-users and the formal routes of communication during the competitive bidding process. Thus co-configuration was beginning to break down.

The delivery architects designed innovative open-plan spaces that aimed to facilitate different styles of learning in line with the educational vision. However, the usability of the built open-plan spaces was severely undermined by pressures on the contractor to meet budgetary requirements during construction once financial close was completed. At this point major contradictions were surfaced.

The decisions made at this stage were critical to whether the initial educational vision of the school could be delivered. There was a continuum of degree of inclusive consultation and participation of all parties in the design and construction processes. At one extreme we found that different team members felt frustrated by the conflicts between their contractual obligations to their client (now the contractor) and their professional obligations to the end-users (the school). In such a situation the delivery architects stressed that there was a notable absence in the form of a quality and design compliance monitor working for the council and the schools during this stage of the project. Someone acting in this role could have ensured the built design would provide educational spaces that could deliver the educational vision set out at the inception of the project. This was the case in schools at the other end of the continuum. An example being a case where the headteacher unofficially became the quality and design compliance monitor and fought hard for a design that would meet her educational vision. This onerous and time-consuming process meant she had to appoint additional staff to fulfil her duties as headteacher. This headteacher acted as a proxy client for the council at a point at which the formal role of client had shifted to the contractor who was, by this stage, driven by motives relating to completion and cost.

It is important to point out that the complex nature of the design process requires active collaboration in the progressive refinement of levels of detail. The preliminary designs were not fixed before the contractors were involved. The council envisaged a process in which ongoing collaboration and mutual shaping in and through time with the active engagement of all parties, which would result in a building that would be deemed to be fit for purpose by all.

Financial pressures on design and construction projects often involve 'value engineering' in which designs are substantially changed to meet budgetary requirements at different points in the construction process. Different patterns of collaboration during value engineering processes can lead to very different outcomes. Where object motives of time and budget held by contractors prevail, then occupiers may be very dissatisfied with outcomes. When occupiers, designers and constructors mutually shape the design to meet the demands of value engineering the satisfactory outcomes are more likely.

We will now provide details of the ways in which individual design and construction projects progressed within this overall framework.

From local authority policy to individual school design

School C1

In School C1, the headteacher (Occupation 1) took on the role quality and design compliance monitor in the process and ensured that the contractors delivered a building that met the requirements of the intended educational vision (Figures 5.5 and 5.6). In the schools we have studied, the alignment of educational vision and resulting building was highest when someone from the staff who occupy the building had taken on this role. That is for co-configuration to actually take place there was a need for client continuity, in some form or another, throughout the process.

In this school there was a very high degree of involvement by the current headteacher (O1) in all the vision, design, construction and occupation phases. When asked in interview whether there was anything

FIGURE 5.5 School C1 design concept
Credit: HKS Architects

Design and practice 99

FIGURE 5.6 School C1 design concept
Credit: HKS Architects

she would change about the design of the building as occupied she answered 'nothing'. She acted as the active facilitator of collaboration throughout the project:

> My vision was to create a twenty-firsst-century learning environment where students can flourish, can work independently to build up their social skills as well as develop their intellect.
>
> *(Headteacher O1 interview)*

The site offers large open spaces on single floors and a 'mixed economy of smaller cellular rooms and breakout rooms. Figure 5.1 shows the typical floor plan of the open-plan teaching zones. The process of occupation was considered very carefully. A 'mock up' of the open learning zone and breakout spaces was constructed in the gymnasium of the old school building:

> We mocked up open-plan learning spaces and learnt how to use them effectively to improve progress for our students. We were preparing a good two years before we moved.
>
> *(Headteacher interview)*

This was used as a testbed for the development of new approaches to teaching and learning.

This process of learning to use new configurations of space (Figures 5.7 and 5.8) continued once the building had been occupied. The headteacher argued the case for conscious leadership of her staff in

FIGURE 5.7 School C1 design in practice
Credit: HKS Architects

FIGURE 5.8 School C1 design in practice
Credit: HKS Architects

learning how to use the space and develop pedagogic practice which transcends what were seen as the inevitable pitfalls of trying to teach a single class in a single room. The headteacher also resisted attempts to reintroduce informal delineation of the large open zones:

> We dedicate one night a week to planning and that planning time is purely open-plan learning support therefore if you're teaching with two other members of staff it's not about what you're going to teach it's about how you're going to teach it in a class of 60 kids or with three groups of 90 or 100 kids.
>
> *(Headteacher interview)*

The contradictions between the teachers' personal histories of pedagogic practice and the vision embedded in the new design were actively resisted by the headteacher. She acted as a strong advocate of the new educational vision of the design.

There were also claims that the changes in design and subsequent changes in practice gave rise to behaviour change. In the former school building the headteacher (formerly the deputy) suggested that the behaviour of students was often quite violent and that the new building has reduced aggression:

> The behaviour has improved dramatically, the standards of teaching and learning, the standards of student, the standards of progress and attainment in our school over the last five years has shot up astronomically.
>
> *(Headteacher interview)*

> The visibility of this design has made a huge impact on the children and their behaviour. I had been at the old school for a very long time and it was quite a tough boys school with lots of fights at lunch times, lots of aggression. If you came here at lunch time, you would have seen a row of boys sat outside my office covered in blood. That was the environment that they were in, there's none of that here and I'm not thinking it's because the boys have changed but it's because they know they can be seen and if you see something, you're there you can deal with it.
>
> *(Headteacher interview)*

It was not only staff who had to learn to use the new building design. There was an explicit approach to 'teaching' the new rules of social order in the new spaces of the building:

> We encourage youngsters to regulate themselves to some extent but there are expectations, there are zone protocols, what they can and can't do in the zones, there are expectations about the best way to learn, you are here to progress, you are here to become better learners . . . a lot of emphasis upon independence.
>
> *(Headteacher interview)*

The timetable and management of the school were designed to promote best use of spaces. The headteacher appears to have a relatively informal charismatic and personalised approach with staff and pupils. The staff enjoy team/carousel teaching, and all the 36 students who were interviewed reported that

102 Design and practice

enjoyed active learning while working in groups in the mixed economy of spaces. Parents were initially cautious but with higher attainments the school is now oversubscribed.

This was the only school which witnessed an explicit and overt attempt to learn how to use the spaces of the design as envisioned. The design, construction and occupation processes were characterised by high levels of collaboration and active participation in which flexible groupings of different professionals experimented and prototyped the emergent final design of the building. This was a form of mutual shaping of the emergent design in which no single actor or agency led the process. In a similar way the occupiers of the building worked together to explore the potential of the design that would afford the forms of pedagogy that they sought to develop. The design and the new form of practice which emerged was an outcome of this ongoing pattern of mutual shaping over time.

School C2

The Local Authority Transformation Team heavily influenced the design of School C, which was a major refurbishment of an old school building supported by headteacher (Occupation 1).

Figures 5.9 and 5.10 show the original design with its commitment to open-plan learning. 'The head [O1] was the driving force behind the educational vision' (Project Architect interview) but creative dialogue and collaboration between the architects and the school were halted and heavily managed after the

FIGURE 5.9 School C2 design concept
Credit: DGS Architects

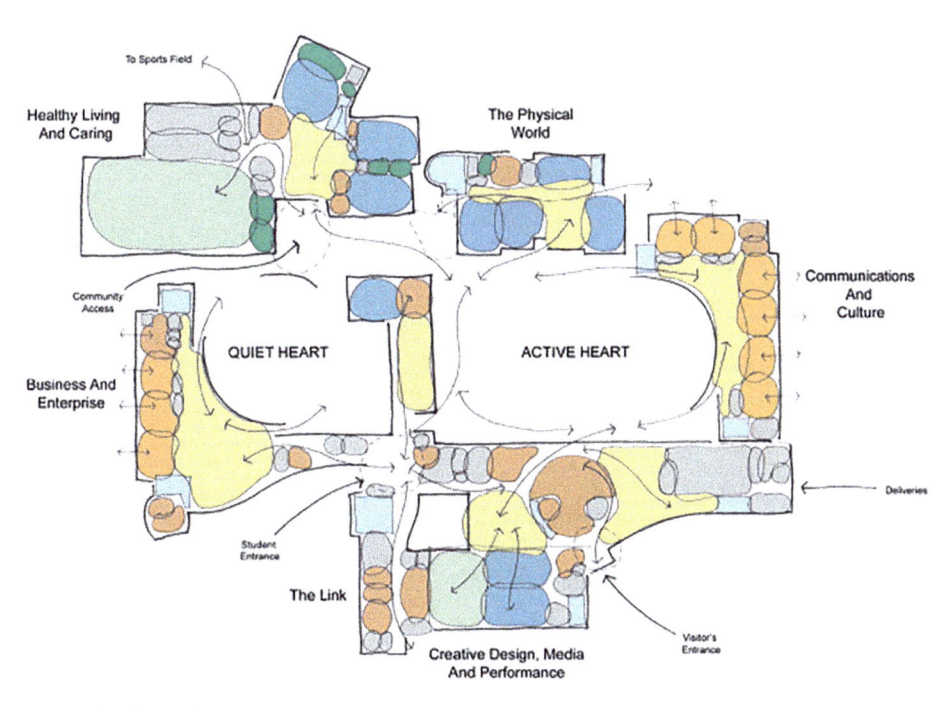

FIGURE 5.10 School C2 design concept
Credit: DGS Architects

contract was awarded to the main contractor. The delivery architects found the competitive process very challenging on resources and engagement:

> it was a frustrating process because at the beginning there were 10 or 11 teams bidding for the work, then that got narrowed down to three so there was all this duplication of effort and particularly for the schools, whenever we had a meeting there could be up to 20 people in the room and you felt that all these other people were just monitoring what was going on and I think the schools, with the costs starting to creep into it, were getting frustrated because they felt their money was being wasted on meetings when actually they just wanted input into the design.
>
> *(Project Architect interview)*

> Once we were on site it was very much controlled by the contractor so our access to the school was very much restrained from that point. Up to winning the bid we were quite involved but once the contractor had secured the project up to financial close there was a barrier between us and the school effectively. I think it was their way of thinking they could control costs that way but that made communication at the end difficult.
>
> *(Project Architect interview)*

> We tried to have regular design meetings with the sub-contractors but there were many instances where we were overruled by the contractor on costs and then consequently the coordination went awry, particularly with steels and duct work impacting on ceiling heights, that was a big issue.
>
> *(Project Architect interview)*

104 Design and practice

Here we have examples of isolated professional action which constituted difficulties and barriers to the development of processes of design work that involved weak forms of collaboration. Communication between the different agencies and agents broke down and the possibilities for the mutual shaping of the outcome were not afforded. Professional isolation of some individuals and passive acceptance by others constituted major problems. From the first occupation by the headteacher contradictions between the design and the preferred practice of staff were clearly apparent. Different object/motives were in play. Unlike School A there was no professional development programme concerned with learning how to use the new design. Staff and students complained that the spaces were noisy and very distracting and also complained that the refurbishment was not fit for purpose. Difficulties in the relations between the leadership and the staff emerged:

> You can have a visionary head but you've got to take your staff along with you as well, there was no prior preparation or involvement for staff and students at the school.
>
> *(Project Architect interview)*

Headteacher (Occupation 2) initiated a programme of wall building to transform the open learning zones into cellular classrooms within eight months of the handover from the contractors to the occupying school staff. These new walls are shown in Figure 5.11. This a physical manifestation of the contradiction

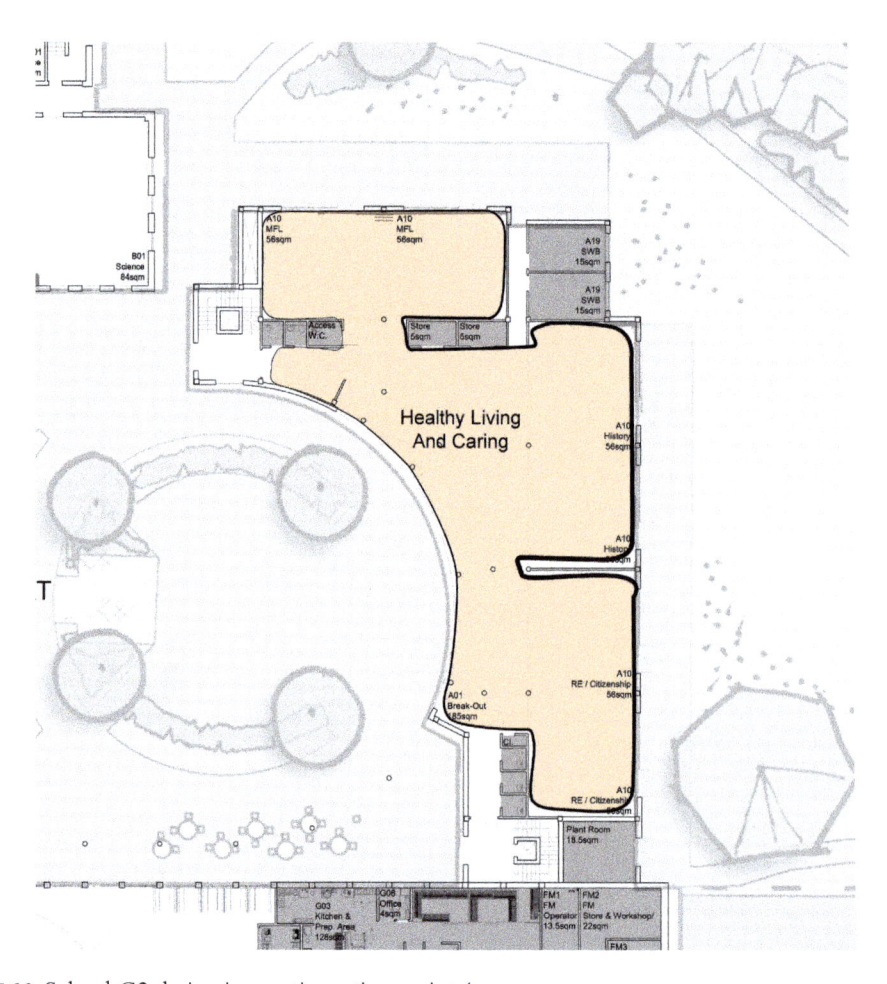

FIGURE 5.11 School C2 design in practice – time point 1

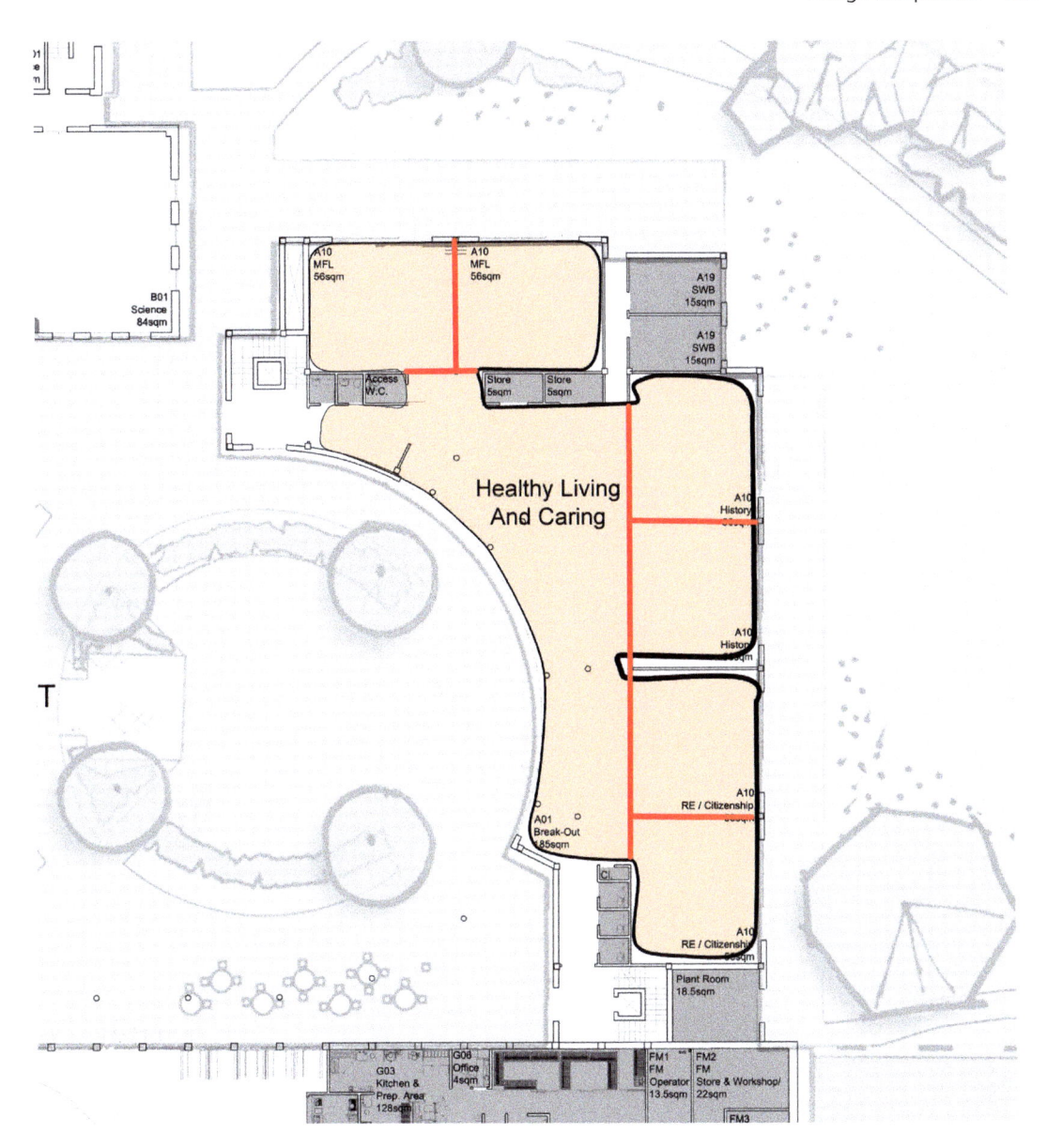

FIGURE 5.11 School C2 design in practice – time point 2

between the vision of the design and the practice developed during the first occupation. Another tension arose with the new appointment of headteacher (O2). Headteacher (O2) was the former deputy at School C1 and is heavily influenced by the head of School C1. The design of the open-plan learning zones was similar to the design of School C1. He expressed concern about the contradictions between design and practice in the previous occupation:

> If you try to deliver traditional style teaching in an open-plan environment, it isn't going to work.
> *(Headteacher O2 interview)*

106 Design and practice

Headteacher (O2) gave a robust defence of the original pedagogic vision developed by the county's Transformation Team. His personal preference would have been to remove the walls that headteacher (O1) had installed. He was very aware that the staff would have reacted badly to such a move. They had complained about the original design and wanted configurations of space that allowed them to continue with their preferred 'single classroom' mode of practice. In the final interview headteacher (O2) also suggested that recent national policy changes had made it more difficult for him to progress his vision for the use of space in the school as it was originally designed. In the short term he had resolved to continue with the modified design pending opportunities for alternative developments.

In the case of School C2, contradictions between object motives of the vision for the refurbishment and the first occupation followed by a return to the original object motives in the second occupation resulted in a legacy residual contradiction.

School C3

This school design project involved a high degree of involvement on the part of the original headteacher (O1) in the vision, design and construction phases. The design was based on the 'schools within a school' model, as shown in Figures 5.12 and 5.13. It is the most radical design of the four new BSF schools in

FIGURE 5.12 School C3 design concept

Credit: HKS Architects

FIGURE 5.13 School C3 design concept
Credit: HKS Architects

this locality. On initial occupation the building had four clusters of 12 open learning zones issuing onto a double height atrium space on the ground floor.

There is also a very large, open 'heart' space that connects all the mini-school clusters. This very dramatic and aesthetically pleasing space is used for dining and whole-school assemblies.

The design assumed integration of curriculum areas. The vision was that of thematic curriculum content taught by teams of teachers who did not 'belong' to departments as much as they did to their 'mini-school team'.

The current temporary headteacher (O2) is developing a much more formal approach to teaching, subject knowledge, departmental structure and discipline. His claim that students, teachers and parents hated the open spaces was used to justify a retrofit. The case was made to the governing body to borrow a large amount of money (£850,000 plus) from the local authority to build glass walls on the front of the open classrooms and to introduce partitions into the open areas within the mini-schools in order to reinstate cellular closed classrooms as shown in Figure 5.6. The new retrofit is claimed to be much more popular with students, teachers and parents. In some cases these were the same families where siblings attended School A and reported high levels of satisfaction with learning in large open zones.

On first occupation in School C3 there was an attempt to align the object motives of the design and the practice (Figures 5.14 and 5.15). There was evidence of partial collaboration in the design and construction process in the absence of the teaching staff. There was no evidence of the development of new discourses and practices of teaching and learning as afforded by the design and were implicit in the original vision for the building. The affiliation to the radical design stemmed largely from headteacher C1 who had not engaged his staff in his vision for the new school. In part the resistance of the staff and

FIGURE 5.14 School C3 design in practice

FIGURE 5.15 School C3 design in practice

perhaps the very radical nature of the design rendered this engagement problematic. The second occupation did not share the pedagogic orientation of the vision and this led to the retrofit.

School C4

In this case, there was little involvement of the original headteacher C4 in the vision, design and construction phases. A deputy headteacher represented the school's views, who has since left the school. We were informed of strong management of the construction phase by the contractor. It was suggested that the constructor kept parties apart and produced a palpable sense of disconnection between educators and other stakeholders. Here there was a deliberate and explicit attempt to prevent active forms of collaboration which from the perspective of the contractor would have inhibited progress towards their desired outcomes:

> I joined the project after we had won the bid, and to me the whole process of developing the design with the school after that seemed wrong. I expected open, informed and dynamic discussions, held within openly stated constraints of affordability. Instead we got only very limited contact with the school, carefully choreographed by the main contractor so that we didn't say a word out of place. The school probably felt they should have had a lot more say in the development of the design than they ended up having.
>
> *(Delivery Architect C4)*

We have gathered accounts of communication being managed between stakeholders with elements of messages being redacted. We also have obtained accounts of the preparation of bids for a contract involving the deliberate obfuscation of limitations in the design with regard to acoustic performance. The design specification including large open spaces on single floors were designed for three groups of pupils of the same age studying the same subject at the same time. The vision was that of three teachers plus classroom assistants working with 90 pupils in the open area and making flexible use of breakout spaces as shown in Figures 5.16 and 5.17.

> Not enough detail was included in the design at both bid submission stage and financial close stage. This meant that the client was very exposed to substandard quality creeping into the design as we detailed it. The main contractor was unwilling to share with us the cost plan during detailed design development, nor use our design expertise in cost-cutting discussions. In the end, they had ignored many of our drawings and specifications and built the way they wanted to, to the detriment of quality and performance.
>
> *(Delivery Architect C4)*

During the design and planning stage significant concerns were reported. The acoustic engineers raised questions about whether the design of open-plan learning models would function effectively. Engineers reported concerns to contractors who were driven by motives concerning securing the contract.

When the contract was won, motives concerning completion and cost deflected attention away from the original acoustic concerns in the value engineering process. Value engineering is a process used to solve problems and identify and eliminate unwanted costs, while improving function and quality. The aim is to increase the value of products, satisfying the product's performance requirements at the lowest possible cost. In construction this involves considering the availability of materials, construction methods, transportation issues, site limitations or restrictions, planning and organisation.

110 Design and practice

FIGURE 5.16 School C4 design concept
Credit: County Council

FIGURE 5.17 School C4 design concept
Credit: County Council

The current headteacher O2 has a strong focus on attainment and has been successful in improving standards. However the school is not managed in a way that aligns with the original educational vision. The timetable does not place same-year groups or subjects in the open areas. The occupation of the building is characterised by informal attempts to change the organisation of space, through placing furniture in such a way as to try to recreate the sense of single classroom spaces. This results in physically awkward spaces which are generally regarded by staff and students as not fit for the purpose for which they are now used as shown in Figures 5.18, 5.19 and 5.20. They are particularly problematic acoustically.

FIGURE 5.18 School C4 design in practice

FIGURE 5.19 School C4 design in practice

112 Design and practice

FIGURE 5.20 School C4 design in practice

However, the pupils state that they generally like the building and that they feel happy and safe in it. They are at ease using the large 'heart space' socially and regard the high visibility within the building as reassuring.

In School C4 there was limited co-configuration in the original design and the first occupation was brief. The second occupation was based on a pedagogic vision which was in stark contradiction to the original design. In the absence of the substantial funding that would have been required to remodel the design, informal approaches to reconfiguring the spaces were invoked. These amplified the disaffection with the original design on the part of staff by virtue of difficulties with acoustics and the intention of the original designers to not allow returns to previous practice. There was no attempt to learn how to use the spaces as designed.

School C5

School C5 was involved in Wave 4 of the county's BSF programme but after two years of school involvement with the vision and design process, the programme was cancelled in 2010 and the project was scrapped. School C5 has been left in the original school building which is in a very bad state of general maintenance and repair. The head who was part of the BSF process feels heavily constrained by the limitations of the traditional design and the small size of the existing cellular classrooms:

> When the decision was made, that there was no money for the school to be re-built, the sole focus was improving pedagogy across the school and there has been a relentless focus on assessment, collaborative learning, engagement and what we tried to do was make lessons more dynamic and

interactive, to demand so much more of our students and to encourage them to actually have extended conversations with each other and with members of staff. But a lot of these activities and a lot of that work requires decent spaces for students to move within the classrooms. I don't know if you have seen many of the rooms but some of them are very old-fashioned rooms and are very small spaces and these spaces do not allow us to do this.

(Headteacher interview)

His approach to meeting the demands for pedagogic transformation established by the local authority would have been to design larger classrooms which would permit a higher degree of flexibility along with the availability of breakout space:

We were really adamant that we didn't want enormous spaces, so that when we were listening to the other schools, yes aesthetically they sounded amazing but we couldn't understand how these would function, particularly if you weren't going to drastically retrain your work force to deliver in these spaces.

(Headteacher interview)

The contradiction between the traditional pedagogic vision and the object motives of a alternative kind of transformation was again stark in this school.

This school went through a satisfactory process of consultation and a design evolved that was mutually shaped to the approval of all parties. This example shows how an alternative set of design assumptions could deliver the same personalisation agenda set out by the council. The council's vision did not determine the design rather it was a component in a complex set of negotiations and discussions. One set of priorities may be interpreted and developed in different ways.

Conclusions

Thus far we have reported some of the findings which reveal some of the ways in which school buildings play a role in mediating the pedagogic process. We have shown how contradictions embedded in this building and the building process help to shape the possibilities for pedagogic practice which in turn may also seek to re-shape the building itself. It is the tensions that are set up between these strands of development which have given us insight into the way in which mediational processes progress after occupation.

Our data suggest that some buildings may be so riven with contradictions that adaptations to particular preferences may prove ineffective and the building becomes seen as dysfunctional. This may either be because of features internal to the design or because of relations between practices of construction and funding. We have evidence of adaptations which were successful in re-shaping these school buildings in a way that rendered them more fit for the purposes of the occupiers.

Our data suggest that the relation between design and practice is crucial to the production of a building which can be and is used effectively. The suitability of the building for a schools' pedagogic practices as they change through time will be determined by the building's potential to adapt to the school's changing spatial needs and the school's understanding of the building's design principles. There are three elements to this relationship. First, it is more likely that a successful occupation and use of a building results when the practices that the occupying staff wish to follow mirror the principles of practice that are embedded in the vision and design. Second, this is most evident when the eventual practitioners (usually the headteacher who takes over the school building on completion) have been involved in an inclusive consultation process throughout the vision, design and construction process. Third, it is quite clear that the principles

114 Design and practice

of the design brief may be regarded differently by different individuals and professional groups. This may seek to compound problems with the relationship between design and practice.

These conclusions lead us to form a general argument that one design may be perceived and used in very different ways in different practices of schooling. We also argue that good design requires good multi-professional holistic post-occupancy evaluation which has a remit that goes far beyond the physical functioning of the building. An understanding of social relations that are enacted within a design as it is taken up by different forms of practice is crucial to the development of better sites for schooling.

In short, we suggest that different patterns of collaboration and mutual shaping of design and practice are vital elements in the processes through which a school is designed, built and occupied. This requires clarity and continuity in the operational definition of the client. In interview a senior architect suggested that the key role of the architect was as 'orchestrator' and a contractor suggested that 'integrity must lie at the heart of a build'. In their different ways they appear to recognise the need for the formation of common objects of the work.

Importantly we have shown how one design can be used in different ways. Rather than design determining behaviour it takes up a dynamic and fluctuating relation with the practices of occupiers resulting in a wide variety of outcomes as illustrated in Figure 5.21 in the production and deployment of the building as an artefact. Different approaches to school leadership and management give rise to distinctive school cultures which in turn make differences in the use and adaption of a school building.

We take the findings as a strong argument for the development of a social and cultural dimension to post-occupancy evaluation which examines human practices in buildings over time and through different management cultures. It is as if there is a process of resignification at each point of cultural change in successive management regimes. As the headteacher of a successful new-build free school noted: 'The design is a provocation to learn differently but it's what you do inside it that matters.'

Taken together these findings point to the need for post-occupancy evaluation that includes human action and perception over time and the inter-connection between design and practice and how this may change over different occupations (school leaders). The concepts of knotworking and co-configuration point to forms of activity which are not easily captured in the term collaboration. In forms of work that are as complex and nuanced as design, construction and occupation there is a need for a more sophisticated model of description that can generate a range of possibilities for collaborative action across agencies,

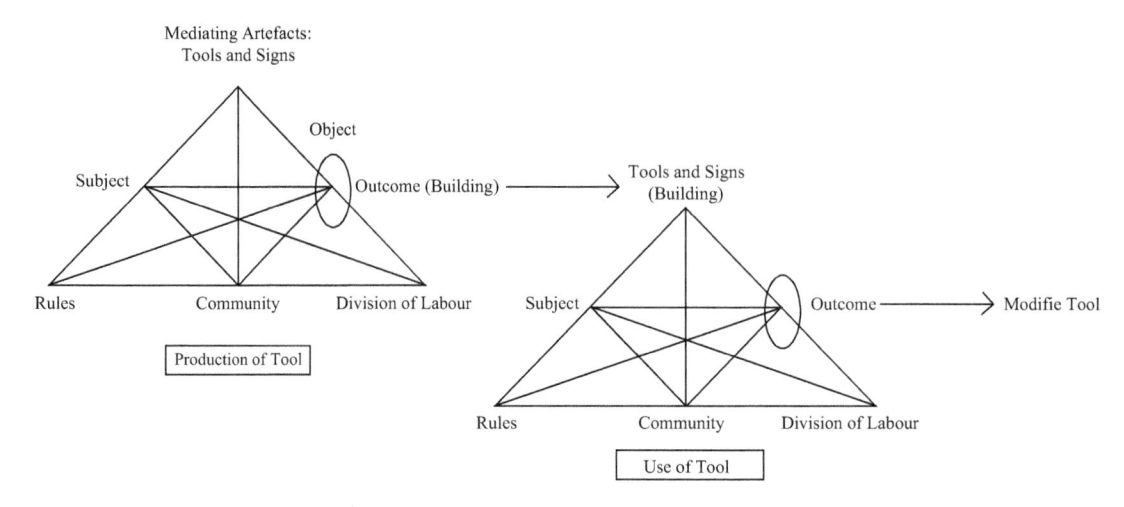

FIGURE 5.21 The production and deployment of the building as an artefact

places and time. The findings also point to the need to redefine 'sustainability' in terms of adaptation to different forms of practice. In order to extend the functional life of new school buildings the vision and design process must allow for adaptation as educational policies and practices change through time (Figure 5.22).

In the terms used by Victor and Boynton (1998) this nuancing of the concept of sustainability calls for a form of work that resembles co-configuration. This calls for a radical rethinking of the work of designers. Rather than designing a building as an event this new form of sustainability requires design as a dynamic practice which enacted through time rather than within a short period of time. The work of designer would not conclude when the building was handed over to the clients rather the designer would maintain a relationship with the client in order that the building can co-configured as the local and national contexts of policy and pedagogic practice shift. These shifts present new landscapes of demand for the design of buildings.

We see similarities here with Waddington's (1956) developmental chreods which were proposed as models of cell development within embryology. In Figure 5.23 the ball represents cell fate. The valleys are the different fates the cell might roll into.

At the beginning of its journey, development is plastic, and a cell can become many fates. However, as development proceeds, certain decisions cannot be reversed. The surface down which the ball rolls is itself subject to change.

However, in Waddington's model this landscape is not fixed, rather it is a constantly shifting landscape which is moulded by differentials in genetic activity as in Figure 5.24

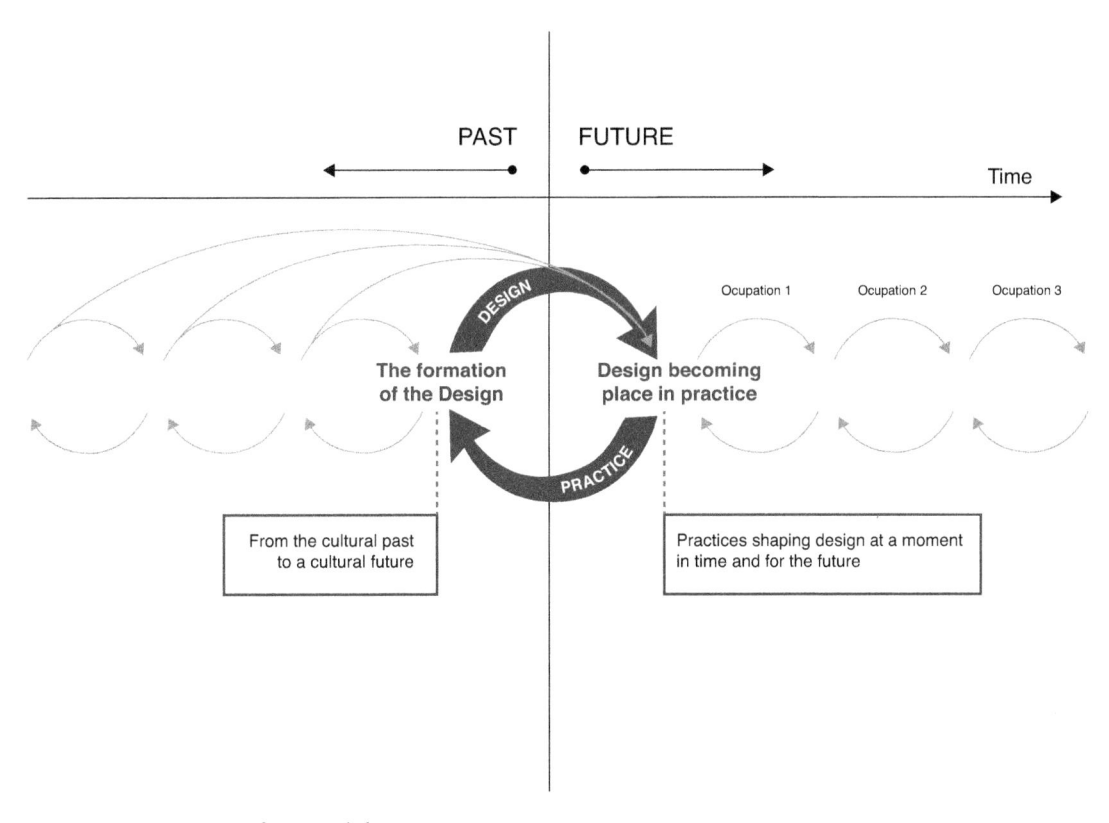

FIGURE 5.22 Diagram of sustainability

116 Design and practice

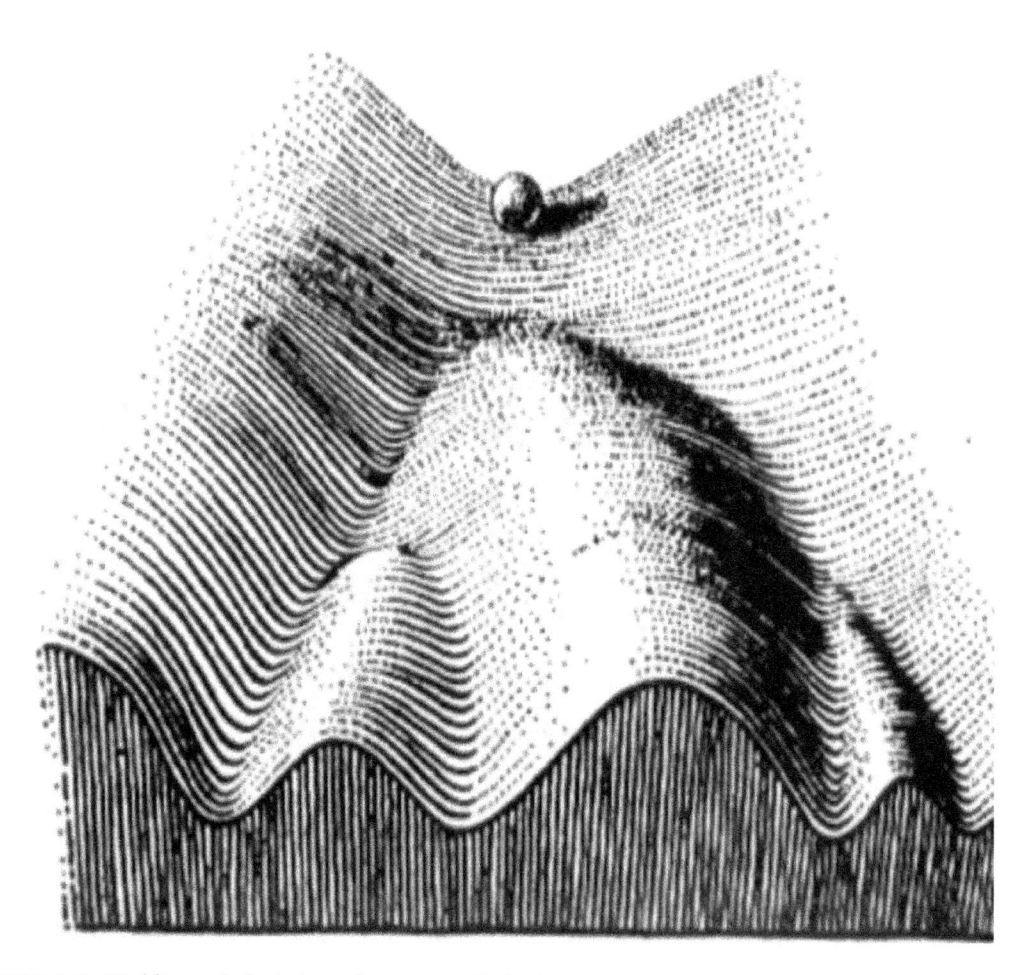

FIGURE 5.23 Waddington's depiction of an epigenetic landscape

This model speaks of constraint yet not in an over-determined manner. As Shotter (1997) noted, it is better to think about the complexity of the fluid, complex, continuously changing landscape of everyday life in terms of a seascape that requires navigational skills. The navigational skills that are required in the ever-changing 'seascape' of educational priorities make demands on the design of buildings. When these demands are unmet, the design may be quickly regarded as dysfunctional or constraining. Ongoing engagement by designers along with all the other stakeholders (e.g. commissioners, clients) in the co-configuration of school buildings in order that they provide the best possible fit to the demands of the often rapidly fluctuating circumstances would enhance the functional life and ultimately the economic value of the original investment in the building asset.

This argument places human action at the heart of the original design process, as the key agent of transformation of the design, and as the key driving force in the progressive adaptation over time. Sustainable buildings are those which are co-configured through a social understanding of how humans interact in learning environments. A first response to our argument about sustainability is that it is not financially

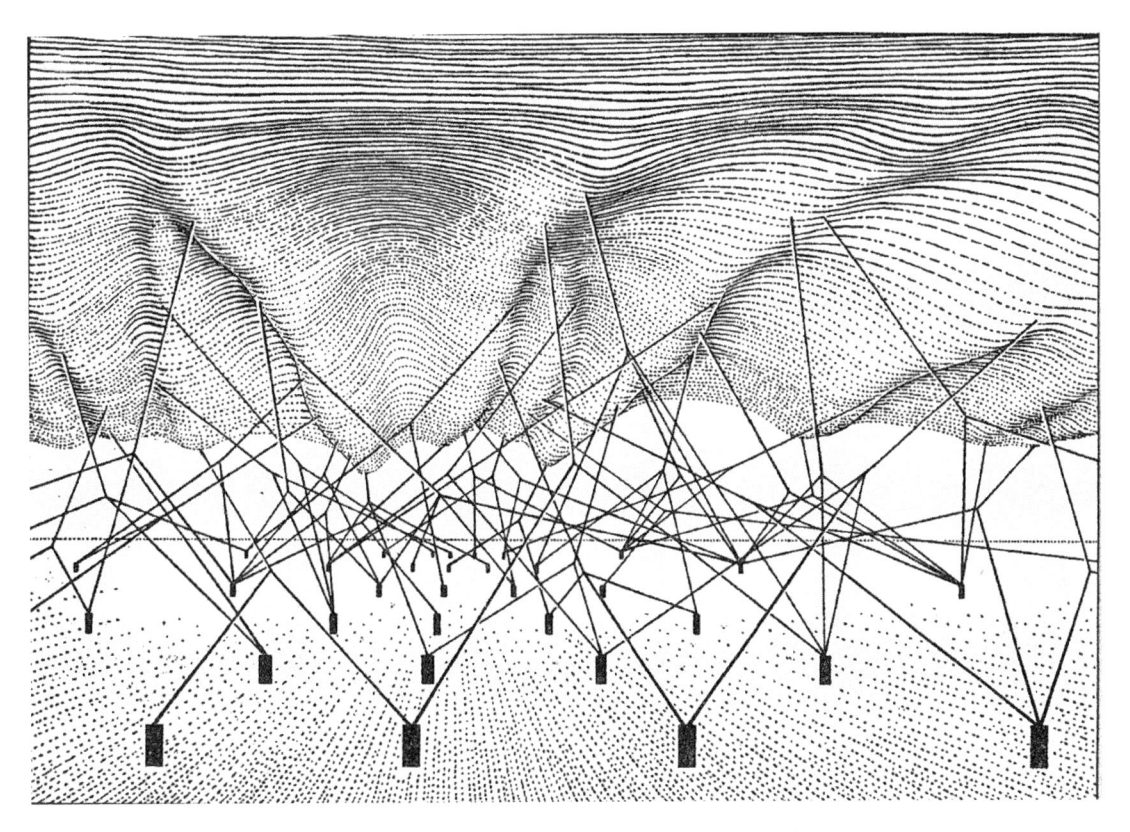

FIGURE 5.24 The flexible underpinning structure in Waddington's epigenetic landscape

viable. Our response is that without responsive designing as a process investments in capital assets may fail to yield the returns which they should provide for investors. Complex forms of action such as those that take place in schools require designs that do not constrain practice rather they should facilitate it.

References

DfES (2003) *Building Schools for the Future. Consultation on a New Approach to Capital Investment.* London: DfES.
Hillier, B. (2007) *Space is the Machine: A Configurational Theory of Architecture.* London: Space Syntax.
Hillier, B. and Hanson, J. (1989) *The Social Logic of Space.* Cambridge: Cambridge University Press
James, S. (2011) *Review of Education Capital.* https://assets.publishing.service.gov.uk/government/uploads/system/uploads/attachment_data/file/180876/DFE-00073-2011.pdf
Milliband, D. (2006) Choice and Voice in Personalised Learning, in *Personalising Education.* Paris: OECD, pp. 21–30.
NAO (2009) The Building Schools for the Future Programme: Renewing the secondary school estate. www.nao.org.uk/report/the-building-schools-for-the-future-programme-renewing-the-secondary-school-estate/
Ratti, C. (2004) Space syntax: Some inconsistencies, *Environment and Planning B: Urban Analytics and City Science*, 31(4), 487–99.
Shotter, J. (1997) Wittgenstein in practice, in C.W. Tolman, F. Cherry, R. van Hezewijk and I. Lubek (eds), *Problems of Theoretical Psychology.* York, Ontario: Captus Press, pp. 27–34.
Victor, B. and Boynton, A. (1998) *Invented Here: Maximizing Your Organization's Internal Growth and Profitability.* Boston, MA: Harvard Business School Press.
Waddington, C.H. (1956) Genetic assimilation of the bithorax phenotype, *Evolution*, 10: 1–13.

Credit: HKS Architects

6
THE EXPERIENCE OF NEW-BUILD SCHOOLS

Introduction

This chapter explores the ways in which design influences the perceptions and actions of students and teachers. An understanding of the relationship between design and pedagogic practices through time extends typical post-occupancy evaluations (POEs) which focus on environmental issues such as acoustics, lighting and temperature, using predominately quantitative methods that often fail to explore how different environmental and social factors interact dynamically with users through time. There is also a lack of attention to the ways in which the processes of occupation may shape the experience of such spaces. This chapter reports on one area of the wider study which involved case study profiling to document a range of key issues experienced by teachers and students at each of these schools. Thus it extends previous evidence on the ways in which habitation alters or rejects original design. This chapter discusses the changing experiences of students and teachers in settings where the mutual shaping of design and practice has taken place over time. These findings contribute to the development of a more holistic understanding of the ways in which design may contribute to processes of pedagogic transformation.

In this chapter we argue that spaces that are designed for specific approaches to teaching and learning may be transformed when these spaces are used in practice. We have evidence that subsequent changes of leadership often involve further modifications of the designs and the practices of teaching and learning. We further suggest that these changes have significant consequences for the everyday experience of schooling as evidenced in the voices and actions of teachers and students.

These issues are of particular importance at the moment as school renewal is a major international concern (OECD, 2010, 2017; Ministry of Education, New Zealand 2011). In England, the National Audit Office (2017) has brought public attention to the parlous state of the school building estate. They point to three concerns: the condition of the school estate; the rising demand for school places; and problems with delivering capital projects. It is clear that we need to learn from the experiences and outcomes of recent approaches to designing and building new schools:

> As the Department recognises, significant challenges remain. The condition of the school estate is expected to worsen as buildings in poor, but not the worst, condition deteriorate further. Pupil numbers are continuing to grow and the demand for places is shifting to secondary schools, where

places are more complex and costly to provide. The Department, local authorities and schools will need to meet these challenges at a time when their capacity to deliver capital programmes is under growing pressure as revenue budgets become tighter.

(NAO, 2017, p. 12)

We present research which examines the relationship between the future of schooling as designed and schooling as it is practised. More specifically, this chapter will discuss the perceptions of students and teachers in schools in which there may or may not have been contradiction between practice as designed and practice as enacted. This adds to the historically well-established understanding that occupation may change over time.

In Chapter 4, we have shown how humans shape their buildings through design practice and how they shape their organisations through management practice as well as how organisations and buildings constrain direct and deflect the attention.

Importantly, we have shown how one design can be used in different ways. Rather than design determining behaviour, it takes up a dynamic and fluctuating relation with the practices of occupiers resulting in a wide variety of outcomes. Different approaches to school leadership and management give rise to distinctive school cultures which in turn make differences in the use and adaption of a school building. We take these findings as a strong argument for the development of a social and cultural dimension to post-occupancy evaluation which examines human practices in buildings over time and through different management cultures. It is as if there is a process of resignification at each point of cultural change in successive pedagogic approaches. This is witnessed in the ways in which practitioners refer to and make use of the building before and after changes that place in pedagogic practice in the school whether these changes are influenced by national, local or institutional influences.

Similarly the data in *Design Matters?* (Daniels et al., 2017) make a strong case that design itself should be analysed as a social practice and is a prime example of multi-agency working both at moments in time and over time. It is a form of work which is driven by multiple motives often with contradictions within and between phases of the overall process.

We also identified important aspects of learning in the overall process. If a building which is fit for purpose is to result, then we have seen that agencies (e.g. engineers, architects) have to learn to work together and with end-users in ways that parallel the demands placed on professionals in other multi-agency context such as child protection (Tse et al., 2014). Similarly we have shown that there is a need for designers and constructors to learn from each project and to bring that learning into the next project rather treating each new build as a prototype (Tse et al., 2014). When a building is handed over to the end-users there is often a need for a programme of learning to be established which enables occupiers to make best use of the building on occupation. Taken together our findings suggest that *change/enhancement* strategies will need to be progressively developed as pedagogic practices change to ensure buildings can adapt to dynamic demands of its users (Daniels et al., 2017; Tse et al., 2014). To extend the functional life of a school building, design and practice have to evolve in a dynamic relationship in order for buildings to continue to be fit for purpose.

Theoretical orientation

Drawing on sociocultural psychology as we consider the relationship between design and practice and the impacts on individuals, in the *Design Matters?* project school buildings are understood as artefacts which mediate pedagogic processes. The sociocultural theorist, Vygotsky (1987) viewed the concept of mediation as being central to his account of social formation of mind. It opens the way for the development of a non-deterministic account in which mediators serve as the means by which the individual acts upon and is acted upon by social, cultural and historical factors in the course of ongoing human activity. We argue that in order to understand processes of mediation it is necessary to take into account ways in which activities are structured by their institutional context (Daniels et al., 2018). Vygotsky attached the greatest importance to the school itself as an institution. His particular interest lay in the structuring of time and space and the related system of social relations (between pupils and teacher, between the pupils themselves, between the school and it surroundings, and so on) (Ivic, 1989).

In a discussion of his conception of cultural psychology which is heavily influenced by Vygotsky's writing, Michael Cole provides an account of the term 'prolepsis' understood as a cultural mechanism that brings the end into the beginning (Cole, 1996, p. 183). In the context of child development he argues that:

> The distribution of cognition in time is traced sequentially into (1) the mother's memory of her past, (2) the mother's imagination of the future of the child and (3) the mother's subsequent behaviour. In this sequence, the ideal aspect of culture is transformed into its material form as the mother and other adults structure the child's experience to be consistent with what they imagine to be the child's future identity.
>
> *(Cole, 1996, p. 185)*

In the context of school design we suggest that an ideal aspect of culture (the theory of pedagogic practice) is transformed into a material form (a building) as the commissioners and architects attempt to structure the teacher's and student's experience of schooling to be consistent with what they imagine to be the future practices of schooling.

However, proleptic instruction also suggests instruction that takes place in anticipation of competence. Thus a learner may be encouraged to participate in an activity which as yet they cannot perform alone. This assumption or anticipation of competence in a social context supports the individual's efforts and encourages the learner to make sense of the situation in a powerful way. As Reid and Stone (1991) note, what is meant is not only determined by the physical context, however, but also depends on the social context of the *adult's intended goal*. Thus, the child is led to infer a new perspective, one that is the *joint product* of the child's own initial perspective and that of the adult. The case of design is radically different from this kind of instructional setting. There is no social context in which children and staff are encouraged to enact the practices of schooling with forms of competence imagined in the design. Most designers often do not have the opportunity to engage with the practitioners once the building has been occupied. The implicit assumption is that the practice will be mediated by the artefact – *the building* which is laden with the new imaginations of practice. This theoretical position suggests that if a design is to change practice then practitioners should be in receipt of the forms of social support or leadership which promotes the imagined form of practice. They have to learn to use the design.

In a way that parallels the literature on teaching and learning theories of determination, Griffin and Cole (1984) challenge the notion of instruction as an activity that determines development:

> Adult wisdom does not provide a teleology for child development. Social organisation and leading activities provide a gap within which the child can develop novel creative analyses.
>
> *(Griffin and Cole, 1984, p. 62)*

In a similar way we are challenging the idea that design determines practice. In a way that parallels Reid and Stone's (1991) argument about teaching, we argue that effect of design depends on the social context of its eventual occupation. As the headteacher of a newly built school remarked 'the design is a provocation to learn differently but it's what you do inside it that matters'.

Pedagogic approach to post-occupancy evaluation

In this section we describe the *Design Matters?* approach to the study of the impact of design in four schools which were part of Wave 3 of the Building Schools for Future (BSF) programme introduced in 2004. *Design Matters?* extended POE analysis by assessing the impacts of contemporary design on users in terms of its implications for educational theory, everyday experiences and pedagogical outcomes. As Kraftl (2006) noted, the 'discourse' of the school can be understood very broadly: 'buildings involve constant, material *work* (as much as inhabitation) for discourses invested in them – such as "childhood" or "education" – to retain their meaning' (Kraftl, 2006, p. 488). The project sought to enhance understanding of the qualitative dimensions of school environments, emphasising the vital importance of perception and use of space by students, teachers, parents and others in the school community, in terms of both what is 'denoted' through design and educational aims and what is 'connoted' in terms of the significance of such spaces for users.

A key distinction between our work and that of many other investigations of post-occupancy evaluation is that we followed young people from one setting (their primary school) into another setting (their secondary school). Students came from a range of primary settings to each secondary school and we were able to scrutinise these transitions at group and individual levels. Additionally we were able to examine the dynamic nature of physical and pedagogic transformation and change in leadership in schools over time. This enabled us to study what may be thought of as trajectories of cycles of artefact mediation over time in which the configuration of the built environment, understood as a tool (Daniels, 2001), progressively mediates the activity of the individuals and groups who occupy the building.

Methodology

This chapter draws on data collected from a subsample of the *Design Matters?* project. This subsample of four secondary schools in Locality D were either newly built or refurbished between 2010 and 2012 using BSF funding. The schools were located within the same school catchment under the guidance of the same Local Authority Transformation Team.

Each school constituted a unit in a series of case studies designed to probe in depth the working practices of practitioners (teachers, headteachers) and students in this subsample of schools. Two cohorts of students in subsequent years of Year 6 in feeder primary schools and Year 7 of the secondary schools participated in this project. In this chapter we report the data captured in individual and small group interviews.

In order to gather data on perceptions of the interactions between the physical environment and pedagogic practice we used a method that is designed to elicit data which are not overly determined by the structure or content of researcher questions. The Nominal Group Technique (NGT) procedure is: 1) individuals generate ideas during or before the group meeting; 2) each person takes a turn reading one of their ideas and ideas are written in a central place until all are listed; 3) the group discusses the ideas, possibly adding ideas to the list; 4) each group member ranks the listed ideas; 5) individual rankings are summarised for each idea to form a group ranking; 6) the group ranking of ideas is discussed (following Chapple and Murphy, 1996). A sub-sample of the student cohorts engaged in NGT tasks in groups of eight to twelve (Chapple and Murphy, 1996). At time points 1 and 2 (end Year 6 and beginning Year 7), we explored what students felt would be, or was, different about life in their new secondary school.

Design Matters?

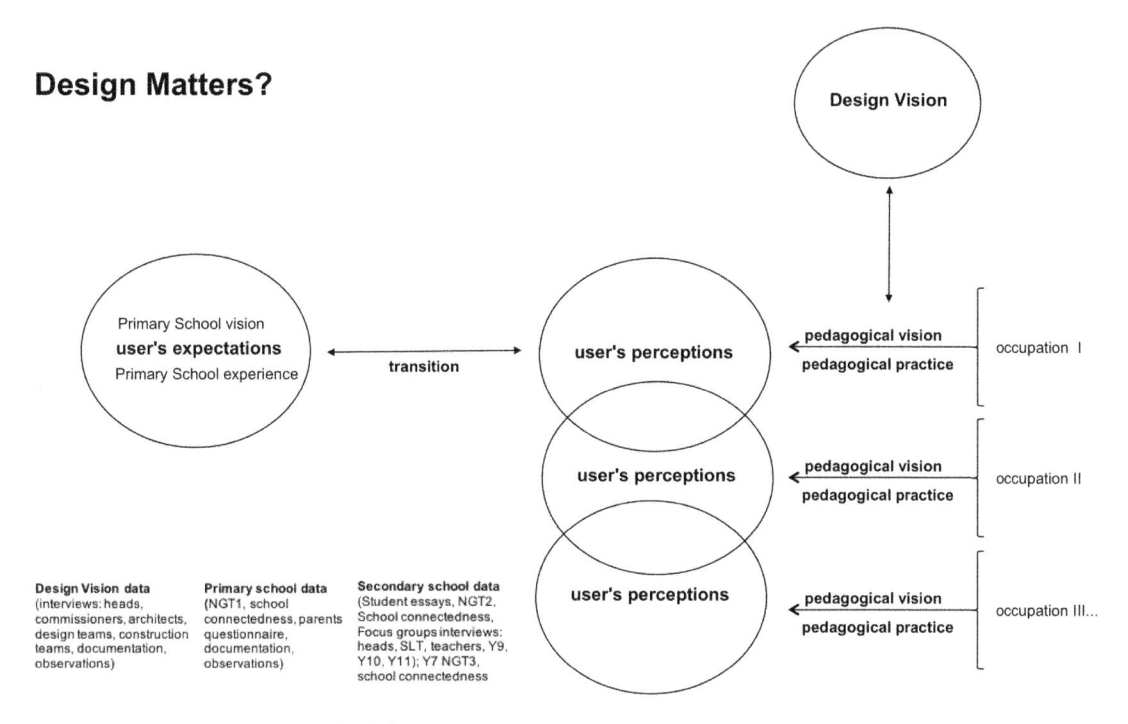

FIGURE 6.1 *Design Matters?* methodology

We also asked students to list the places they felt were most important and/or most enjoyable, and their opposites, and elicited the discussion from these initial lists. In this chapter we have drawn on the students' statements made in NGT settings in order to explore their perceptions of their school environments.

As can be seen in Figure 6.1, perceptions data were gathered at primary school and at each occupation (new headteacher) of the secondary school. Alongside these perceptions data we observed the use of space in each school environment. These observations in turn informed the questions that were used in individual and group interviews. For example, if we observed differentials in the usage in the communal 'heart spaces' of the school, we spoke with students and teachers about their perceptions of these spaces and their explanations for the ways in which they were used. We progressively refined our understanding of use and meaning of spaces through cycles of observation and interview. We not only progressively focused on emergent issues but also followed students from their primary setting through any subsequent transformations of their secondary setting as changes in leadership were invoked by successive headteachers.

Perceptions of the school environment

In an earlier paper (Daniels et al., 2018), we described how the designs of four case study schools were developed from one educational vision of one local authority. The findings showed how *one* design can be used in different ways. In this chapter we describe how these four buildings are *experienced* by the students, teachers and headteachers of the schools and how *one* design may be perceived in different ways, subject to the relation between design and practice in occupation. These data all relate to perceptions of social belonging, and connectedness gathered at different time points during transitions from primary to

124 The experience of new-build schools

Sample
"What do you think will be different about your new school?"

- Personal wellbeing (emotions, belonging) **2%**
- Place/Space (size, navigation, design) **23%**
- Social relations (students, teachers) **35%**
- Logistics (uniform, travel, lockers) **15%**
- Academic (class, homework) **25%**

Comparator
"What do you think will be different about your new school?"

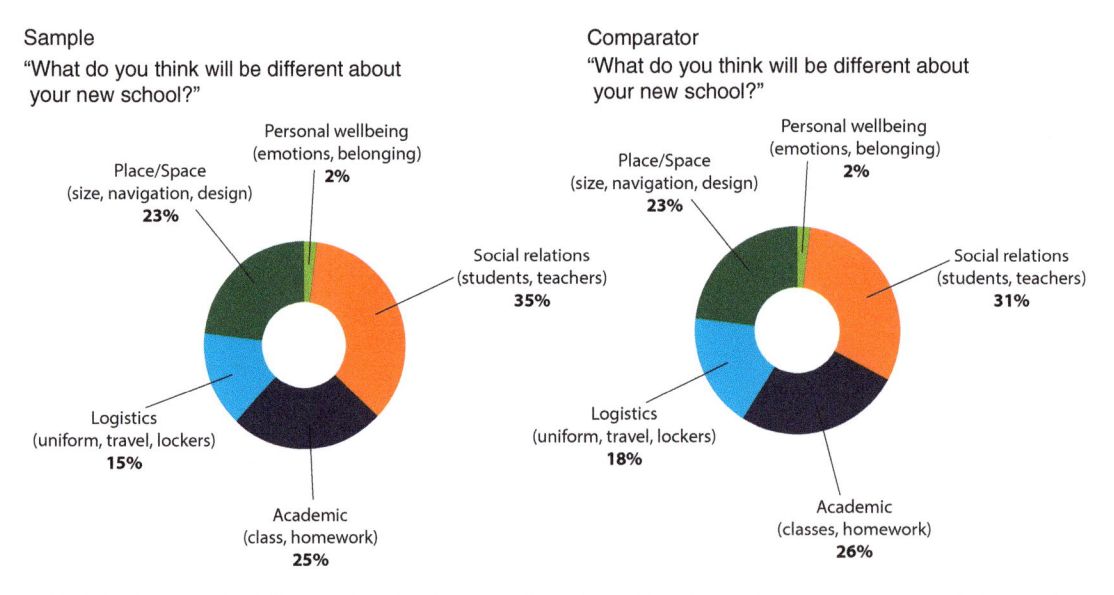

- Personal wellbeing (emotions, belonging) **2%**
- Place/Space (size, navigation, design) **23%**
- Social relations (students, teachers) **31%**
- Logistics (uniform, travel, lockers) **18%**
- Academic (classes, homework) **26%**

FIGURE 6.2 Time point 1: Proportion of primary students that referred to each category, by type of school when asked 'What do you think will be different about your new school?'

secondary school. Importantly, these data enable us to gain a view of the implications of placement in different types of school and transitions between different types of school design. These data provide an insight into the subjective experience of school design in practice.

In Figure 6.2, the diagram shows how all students both in new-build and comparator primary schools refer significantly more often to 'social relations' than other categories when asked 'What do you think will be different about your new school?' Students (end of Year 6) about to transition into secondary schools have most concerns about changes to their social relations; for example, new friends, new teachers, missing old friends. A Chi-square test revealed that these differences were significant at the 0.05 level ($x2$ (1) = 4.01, p < .05).

The proportions of the categories of students' perceived concerns prior to transition showed no significant differences between students going to new-build and comparator secondary schools.

Figure 6.3 shows how students on entry (start of Y7) to their new secondary school responded to the question when asked 'List in order of priority, the most important spaces to you?' Students in new-build schools refer significantly more often to 'social spaces' than other categories and students in comparator schools refer significantly more often to 'academic spaces' than other categories.

In Figure 6.4, the diagrams show how students' responses to preferences for categories changed through time in School B1. From time point 1 to 3, students increasingly refer significantly more often to 'social spaces' than other categories. Through interviews and observation data (time point 2 to 3), students stated that school rules do not restrict students' use of social spaces during break times and allow free movement around the site.

In Figure 6.5, the diagrams show how students' responses to preferences for categories changed through time in School D1. In time point 2, students refer significantly more often to 'social spaces' than other categories; in time point 3, students refer less often to 'social spaces' than other categories. Through interviews and observation data (time point 2 and 3), students stated that school rules progressively restricted student's use of social spaces, break times and free movement around the site.

In Figure 6.6, the diagrams show how students' responses to preferences for categories changed through time in School A1. In time 1, students refer significantly more often to 'social spaces' than other

Sample

- Personal well-being (pastoral spaces: SEND, tutor rooms) **9%**
- Logistics (lockers) **2%**
- Social relations (social spaces: public/private, active/passive) **49%**
- Academic (academic spaces: formal/informal, active/passive) **40%**

Comparator

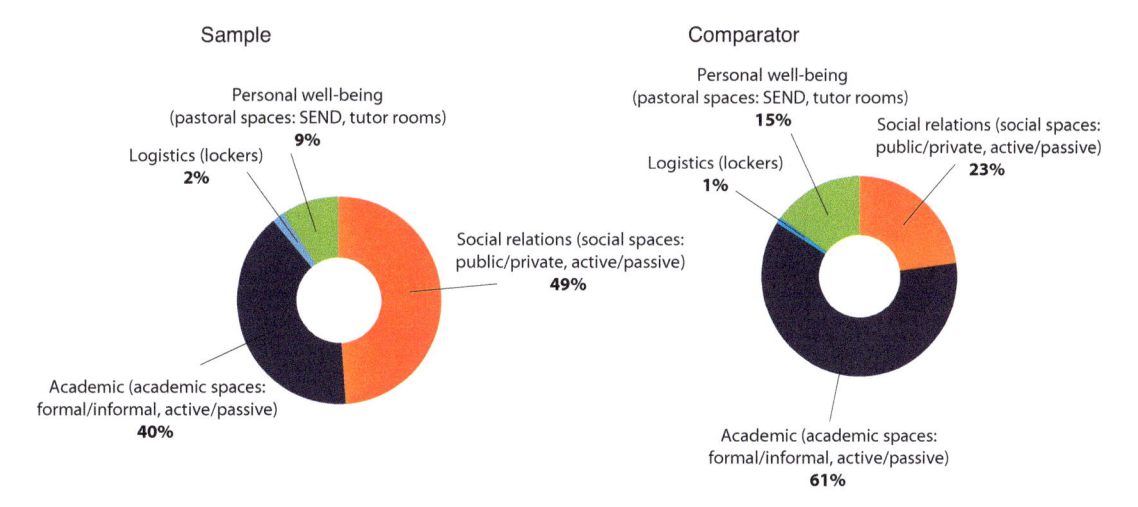

- Personal well-being (pastoral spaces: SEND, tutor rooms) **15%**
- Logistics (lockers) **1%**
- Social relations (social spaces: public/private, active/passive) **23%**
- Academic (academic spaces: formal/informal, active/passive) **61%**

FIGURE 6.3 Time point 2: Proportion of secondary students that referred to each category, by type of school when asked 'List in order of priority, the most important spaces to you?'

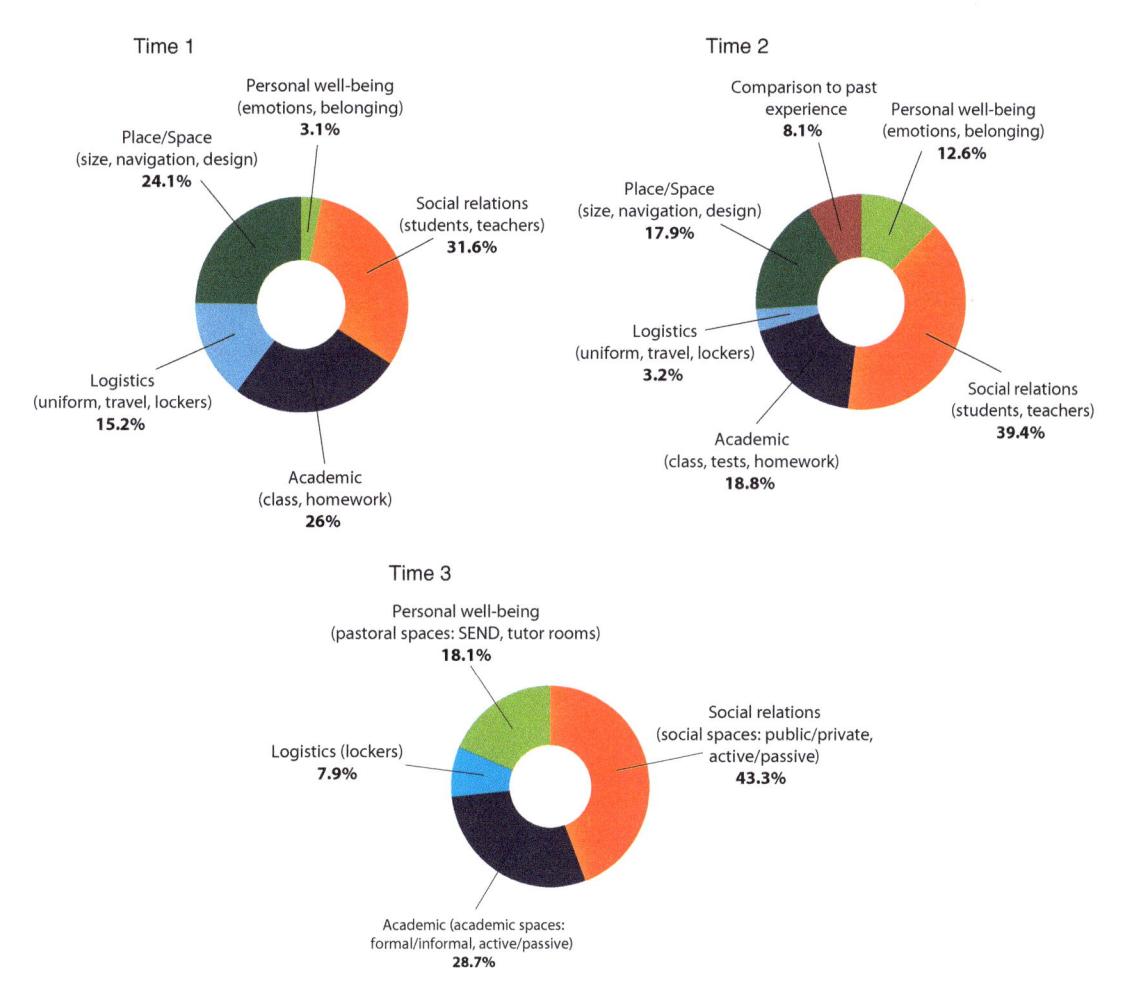

Time 1

- Personal well-being (emotions, belonging) **3.1%**
- Place/Space (size, navigation, design) **24.1%**
- Social relations (students, teachers) **31.6%**
- Logistics (uniform, travel, lockers) **15.2%**
- Academic (class, homework) **26%**

Time 2

- Comparison to past experience **8.1%**
- Personal well-being (emotions, belonging) **12.6%**
- Place/Space (size, navigation, design) **17.9%**
- Logistics (uniform, travel, lockers) **3.2%**
- Social relations (students, teachers) **39.4%**
- Academic (class, tests, homework) **18.8%**

Time 3

- Personal well-being (pastoral spaces: SEND, tutor rooms) **18.1%**
- Logistics (lockers) **7.9%**
- Social relations (social spaces: public/private, active/passive) **43.3%**
- Academic (academic spaces: formal/informal, active/passive) **28.7%**

FIGURE 6.4 Time point 1–3: How secondary students' responses to preferences for categories changed through time in School B1

Time 1

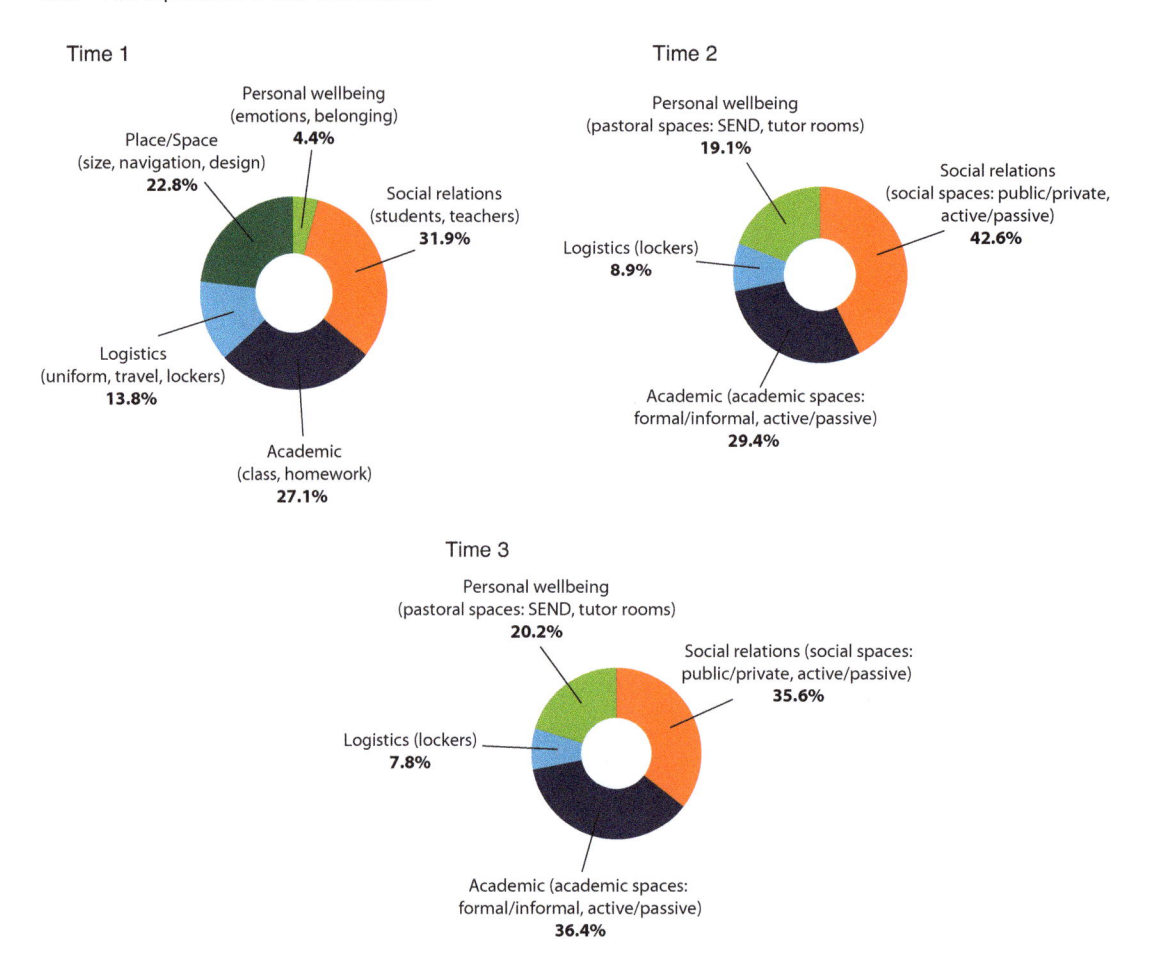

- Personal wellbeing (emotions, belonging) **4.4%**
- Place/Space (size, navigation, design) **22.8%**
- Social relations (students, teachers) **31.9%**
- Logistics (uniform, travel, lockers) **13.8%**
- Academic (class, homework) **27.1%**

Time 2

- Personal wellbeing (pastoral spaces: SEND, tutor rooms) **19.1%**
- Social relations (social spaces: public/private, active/passive) **42.6%**
- Logistics (lockers) **8.9%**
- Academic (academic spaces: formal/informal, active/passive) **29.4%**

Time 3

- Personal wellbeing (pastoral spaces: SEND, tutor rooms) **20.2%**
- Social relations (social spaces: public/private, active/passive) **35.6%**
- Logistics (lockers) **7.8%**
- Academic (academic spaces: formal/informal, active/passive) **36.4%**

FIGURE 6.5 Time point 1–3: How secondary students' responses to preferences for categories changed through time in School D1

categories. In time point 2, students refer even more often to 'social spaces' than other categories. From time point 3 to 4, students progressively refer less often to 'social spaces' than other categories. Through documentary evidence, interviews and observation data (time point 2 to 4), students stated that schools rules progressively restricted student's use of social spaces, break times and free movement around the site. Plans show the construction of physical barriers within the building from time point 1 to 4 and the progressive divergence between the educational vision for the design and the educational practice enacted through time point 1 to 4.

The data points to the importance of social relations and social spaces in schools for students. However, in NGT interviews students stated that the provision of spaces designated as social does not inevitably lead to their use as designed. This is witnessed in the following example of two schools with very similar designs which included a central 'heart space' designed to promote social exchange where *one* design is experienced in very different ways.

In one school the leadership argued in favour of high degrees of visibility in the design in terms of the promotion of a sense of community and belonging. In the other school the leadership extolled the virtues of high visibility in terms of passive control and surveillance. In the latter, students did not use the 'heart' space because they felt exposed and vulnerable, whereas in the former, the central 'heart' space was full of activity during class and break times, students reported enjoyment and a sense of ease in occupying the

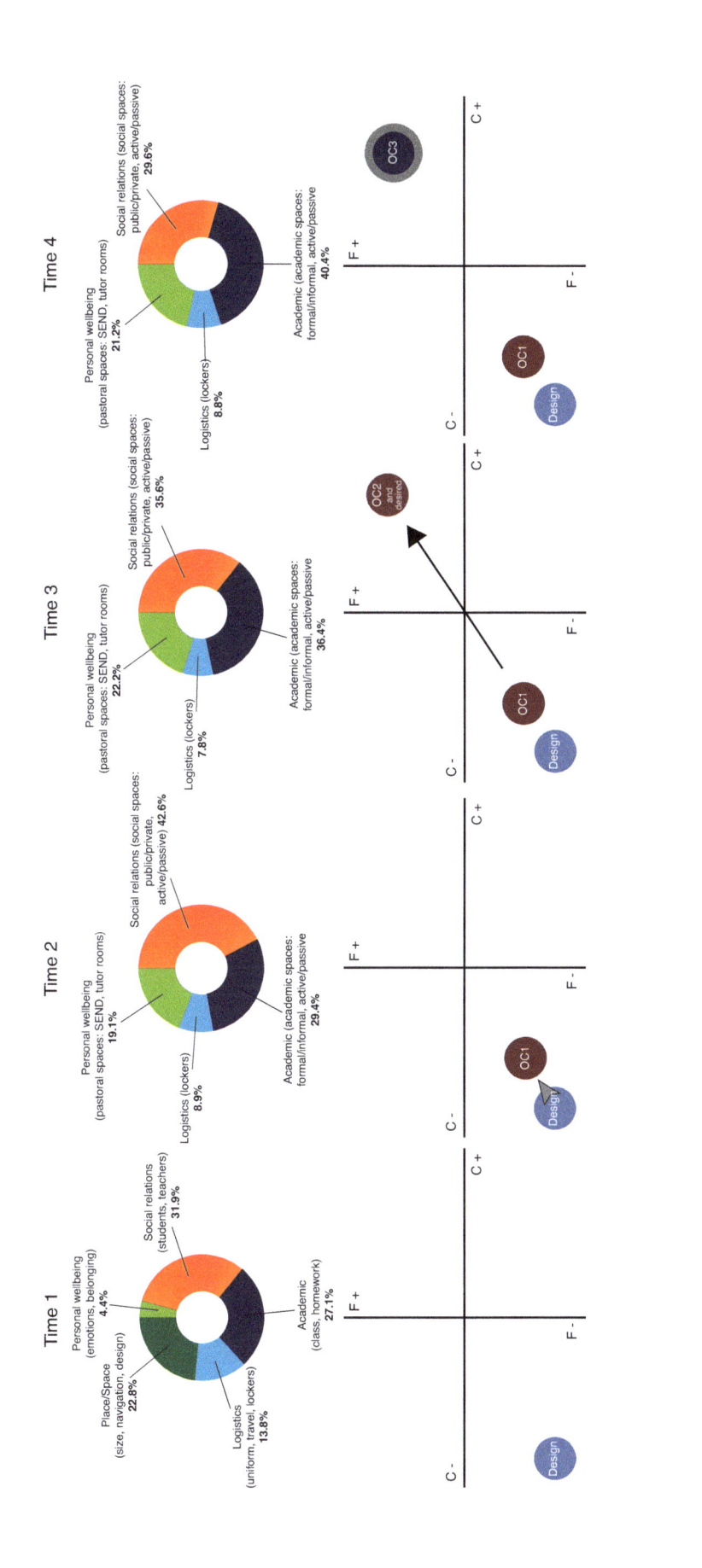

FIGURE 6.6 Time point 1–4: How secondary students' responses to preferences for categories changed through time in School A1 as the design and educational practices changed

space. The students felt safe due to the high level of visibility in the design and did not report concerns about being observed:

> I met all my mates and my besties in the heart space like usual. The heart space is really open, where you can eat, hang out with friends. I love that you can see everything that's going on.
> *(School B1, student NGT)*

> I feel like an ant because we have to walk in a line across the open heart space. It feels cramped and people can see you from all around. We are not allowed to be there at break time.
> *(School D1, student NGT)*

This attests to an important interface between design, the practices of leadership and the experience of space. We now move to an examination of the details of each of the four case studies in our subsample as witnessed in the teachers' interview data and students' statements made in the NGT workshops. This section of the chapter will illustrate the ways in different pedagogic practices give rise to differences in personal experiences of design.

Discussion

Our data suggest that social spaces are important for all students wherever they are. However, there appears to be an effect of design and practice on these perceptions.

Where design was aligned with practice in school which offered social spaces for learning and association, then the perceptions data show that students felt that social spaces were very important. However, when the same social spaces were occupied in practices which privileged instructional features of schooling, then those same social spaces were not perceived as being important. In a sense the design alone did not determine a sense of belonging or connectedness, rather it is the design in practice which has this effect. Our data make this point quite strongly in that we can examine the impacts on perceptions of belonging and connectedness over time and through different configurations of space. We would contend that these data point to a formative effect of juxtapositions of designs and practice on emergent identities (following Holland and Lachicotte, 2007) or possibilities for identity.

We agree with Holland and Lachicotte (2007) that identities are simultaneously social products (collectively developed and imagined social categories) and personal formations in practice (self-meanings developed through a sociogenetic process that entails active internalisation). Identities are also symbolic (identities are mediated by signs and symbols), reflexive (identities are involved in recognising the self-in-practice and the self as a person) and a source of motivation for action.

We suggest that some configurations of design and practice offer collectively developed and imagined social categories of collaboration and flexibility in which personal formations of collective identity in social spaces become a possibility (not determined). However, in configurations of design and practice which are predicated on priorities concerned with individual attainment, in some cases at the expense of collective endeavour and concerns for well-being, then personal formations of identity in social space are downplayed.

Following Lefebrve, we suggest that these data point to the collective production of space in which neither design nor practice alone predominate. Space is formed in design through human action with design.

More work is needed for us to be able to understand the different forms of collaboration and patterns of perception of belonging and connectedness that may arise in different co-productions of spaces. We do not wish to convey a sense of a bipolar construct of design in practice rather we point to a complex set of relations between design and practice which may be generative of a multitude of consequences.

This research shows a need for a better understanding of the interrelationship between the social structures and the spaces they inhabit. How do school spaces evoke perceptions of connectedness, safety and belonging? Research on school exclusion show that it is often the breakdown of social relationships in schools that triggers the chain of events that can eventually lead to the permanent exclusion of school students (Eva Pomeroy, 2000). In England, exclusion rates have been increasing since (DfE, 2018).

It can be argued that architecture should be re-conceptionised as a tool to support social interactions, if design can promote positive social relations, we need to better understand the processes that underlies the mutual shaping of architecture and the social relations that inhabit the spaces created.

Case studies

School C1

In School C1, the headteacher (C1) led the development of the educational vision from the start of the project and subsequently took on the role of the quality and design compliance monitor in the process to ensure that the design team and contractors delivered a building that met the requirements of the intended educational vision:

> It all began by us really thinking about what we saw learning looking like in the twenty-first century, what sort of building would you need to facilitate that.
>
> *(Headteacher C1 interview)*

The design has five zones of highly glazed teaching spaces which wrap around a central green 'heart' space where the main circulation spaces and movement are located. The deputy head discussed the importance of the form and visibility of the design on how it has changed the way the school community relate to each other:

> Everything works in a circular way, so you are always encompassing each other, encompassing the learning. The students are always looking towards each other and can feel a part of one big community.
>
> *(Deputy headteacher C1 interview)*

The headteacher commented on dramatic improvements in behaviour since the occupation of the new school building. She attributes this to the high level of visibility in the new design which was absent in the old school:

> I'd been at the old school for a very long time and it was quite a tough boys school with lots of fights at lunch times, lots of aggression you know if you came here at lunch time you would have seen a row of boys sat outside my office covered in blood, yeah, that was the environment that they were in, there's none of that here and I'm not thinking it's because the boys have changed but it's because they know they can be seen and actually if you see something, you're there you can deal with it.
>
> *(Headteacher interview)*

Students were inducted into the expectations associated with new forms of practices in the newly designed spaces. The deputy headteacher referred to protocols and expectations for students in the new open learning zones:

> There are expectations, there are zone protocols. We want to encourage youngsters to regulate themselves to some extent . . . we have a lot of emphasis upon independence in this school . . . we've put assessment into place that is linked towards independent learning as well.
>
> *(Deputy headteacher interview)*

The students have responded to the demands of the new forms of pedagogic competence:

> In our old building, we had cramped classrooms and dark long corridors. It felt gloomy and depressing. When we came to this school, it just felt modern. There are five big zones where each space can fit four classes. We have open learning and everything changed.
>
> *(L6th student interview)*

They also identified specific benefits in terms of use of space:

> I feel like I can learn better in a zone than in a classroom. In a zone, you have more space and you can sit with people you work well with. You can move around and get help from each other and different teachers.
>
> *(Y7 student interview)*

The headteacher could also point to objective measures of improved performance both in terms of use of space and academic outcomes:

> Design does matter, the conversations, the way the zones are set up, the way students engage with each other. Since we have had this new building, results have improved enormously. We have more students stay on, we have more students go to university. This building makes them feel wanted and gives them an aspiration.
>
> *(Headteacher interview)*

The headteacher pointed to how the new design allows the school to fully develop a different pedagogic approach that was not possible due to the physical constraints of the old building. The open zones allow students and teachers to group and re-group in flexible ways:

> The design allows us to have the flexibility to teach in large groups in the open zones and take smaller groups into adjacent classrooms that might need a smaller quieter space. We get a lot of support in this school to develop the skills you need for team teaching. It has taken some time to develop the best approach to teach groups of 100+ students as it was so different in my last school.
>
> *(Teacher interview)*

The resulting building matched the educational vision of the occupier and there was a continuity of leadership throughout the processes of vision, design, construction and occupation of the new building.

More importantly, an explicit and overt attempt to learn how to use the spaces of the design as envisioned was witnessed. This was much more than the tacit mediation of the design as artefact. Here we observed explicit mediation of the practice by the headteacher in the social situation of the practice. The timetable and management of the school was designed to promote best use of spaces and a 'mock up' of the open learning zone and breakout spaces was constructed in the old building, as a test bed for the development of new approaches to teaching and learning to prepare teachers and students for occupation of the new design. This process of learning continued once the building had been occupied with Thursday evening meetings for all teaching staff to plan and develop effective pedagogy for shared group teaching. To some large extent these actions countered the problems we discussed with respect to proleptic instruction mediated by design artefacts that there might not be a social context in which children and staff are encouraged to enact the practices of schooling with forms of competence imagined in the design. In this school there was an ongoing form of preparation for new forms of competence.

School C2

School C2 involved a major refurbishment of an old school building and the project was led by the Local Authority Transformation Team. The design was driven by the same educational vision as school A and supported by headteacher C1.

'The head was the driving force behind the educational vision' (Project Architect C2 interview) but creative dialogue and collaboration between the architects and the school were halted and heavily managed after the contract was awarded to the main contractor.

From the first occupation by headteacher C2 contradictions between the design and the preferred practice of staff were clearly apparent. Unlike School A there was no professional development programme concerned with learning how to use the new design. Staff and students complained that the spaces were noisy and very distracting and also firmly stated that the refurbishment was not fit for purpose. Headteacher C2 initiated a programme of wall building to transform the open learning zones into cellular classrooms within eight months of the handover from the contractors to the occupying school staff. This physical manifestation of the contradiction between the vision of the design and the practice developed during the first occupation. An additional tension arose with the appointment of headteacher C2 (the second headteacher at school C2). Headteacher C2 was the former deputy at school C1 and was heavily influenced by the pedagogic design and methods led by the head of school C1. The original design of the open-plan learning zones at School C2 were similar to the design at School A. Headteacher C2 expressed concern about the contradictions between design and practice in the previous occupation:

> If you try to deliver traditional-style teaching in an open-plan environment, it isn't going to work. The teachers are still fairly traditional pedagogically, they are doing a good job but they're not making the most of opportunities that this building can offer.
>
> *(Headteacher C2 interview)*

Headteacher C2 articulated a strong pedagogic argument for the mixed economy of open learning spaces and enclosed classroom spaces that was witnessed in the refurbishment before the new walls were introduced. He recognised that there was a need to convince parents and teachers of the pedagogic benefits of a mixed economy of spaces:

> What we're developing here is a very different model and the example I use to explain to the parents is when you go into schools like this you have to develop a different pedagogy which takes advantage of the spaces that you have.

> Each teacher has to manage a broad range of abilities in a traditional classroom of 30 students, when you combine those classes into an open-plan area, it gives greater flexibility to take smaller groups off and tailor the teaching to different needs.
>
> *(Headteacher C2 interview)*

He was particularly concerned with the benefits of larger flexible groupings of students in which need could be more closely aligned with provision:

> When you initially walk in you might see 80 students and think crumbs that's not very personal. But when you look into the pedagogy behind it, you actually have a team of teachers who know that these students here didn't understand X so I'm going to take them off. You end up with a dissolution of the class that looks like one big class but the phrase we used was called flexible setting so you're developing a pedagogy that was actually very close to the need and really closely matched to individuals.
>
> *(Headteacher C2 interview)*

However, he also recognised that he faced a challenge in convincing his colleagues of the benefits of the original design:

> It's all very well having a vision but you've got to take your staff with you. I would like to take the walls down again and use the opportunities the design offers but we are not there yet. We also need to develop a staff that can work effectively together.
>
> *(Headteacher C2 interview)*

Here again these perceptions can be interpreted in terms of proleptic instruction. Given the lack of preparation for new forms of practice there was no social context in which children and staff are encouraged to enact the practices of schooling with forms of competence imagined in the design. In this school there was no ongoing form of preparation for new forms of competence. This had consequences in terms of teachers' attitudes to the design which stood in stark contrast to those witnessed in school C1:

> I found it so difficult to teach in the open zones, it was noisy and the students were so distracted. It's been much better since the walls were built.
>
> *(Teacher interview)*

However, the students spoke with enthusiasm about the original design and dislike of the spaces that resulted from wall building:

> I liked the open classrooms because it was bright, open and spacious. Now the closed classrooms are so cramped. They are the spaces I least like learning in because you get hot and stressed easily.
>
> *(Y8 student interview)*

The lack of consultation with the original designers carried implications for the size and environmental conditions of the spaces that resulted from the adaptations which are now considered not fit for purpose by students.

However, practice in school C1 witnessed effective and ongoing forms of social support and leadership which promoted the form of activity imagined by the designers. Participants had learned to use the design. In School C2 there was a direct conflict between the legacy practice of the staff and the imagined practice of the design. The proleptic instruction of the designers failed to make an impact on practice. The social conditions for this process to have happened were not in place in the school.

School C3

School C3 is the most radical in its pedagogic vision and design initiated by the county's School Transformation Team. The intention was to deliver personalised learning using a 'schools within a school' model. The pedagogic argument that lay behind the design was for a thematic rather than a subject (discipline)-based curriculum which would be taught by teachers working in teams, with students moving between large open spaces and smaller breakout spaces as they pursued a personalised pathway through the curriculum. On initial occupation the building had four clusters of 12 open learning zones opening onto small double height atrium spaces on the ground floor.

The current temporary headteacher (C3) is developing a much more formal approach to teaching, subject knowledge, departmental structure and discipline. His claim that students, teachers and parents disliked the open spaces was used to justify a retrofit:

> When we came here we were looking at some of the practices that were going on and we had open-plan learning with no walls, groups of 60, cross-curricular, mixed ability and it really wasn't working and the staff were on their knees. Educationally it was crazy to think that it would work in terms of the sound and what you were trying to do. So we spent a lot of time looking at the data and talking to staff and children and the children were very clear, the first thing they said to us was we need walls in this school. You've got to think and accept that education is not going to change radically over the next 100 years so much so that you're not going to need classrooms anymore.
>
> *(Headteacher C3 interview)*

The case was made to the governing body to borrow a large amount of money (£850,000) from the local authority to build glass walls on the front of the open classrooms and to introduce partitions into the open areas within the mini-schools in order to build cellular closed classrooms. The new headteacher claimed that the retrofit is much more popular with students, teachers and parents; however, our student NGT data suggests that the retrofit changed the pedagogic practices and the physical environment in ways that were perceived negatively. In some cases these were the same families where siblings attended School A and reported high levels of satisfaction with learning in large open zones.

Members of staff felt that the new design had interrupted their established form of practice. The design presented them with challenges that they were ill prepared for:

> I had been in the old school for eight years and had to leave after nine months in the new building. It was a disaster, I totally disagreed with the way we were being asked to teach in the open-plan learning zones. The noise was horrendous and the students could not concentrate. I came straight back after the new walls were built, the school works much better and the students are happier.
>
> *(Teacher interview)*

The adaptation was not anticipated in the imagined practice and the architects were not involved in the retrofit exercise. Similar to School B, the environmental conditions (acoustics, ventilation and temperature) created by the retrofit have created new problems for teachers and students:

> I hate the closed classrooms, they are so hot, I feel like I can't breathe sometimes. I wish you could spread yourself out and everything rather than being in a cramped space.
>
> *(Y8 student interview)*

> It's worse now, because it was open and everyone's around. They didn't want to shout as much but when it's closed no one can hear them from outside. Now they shout louder and we are all cramped in this small space.
>
> *(Y8 student interview)*

The lack of preparation for participation in new forms of practice envisaged in the original design resulted in the design being understood as an unwelcome imposition that was resisted or dismissed by the second wave of school leadership. These preferences were not shared by the school leadership and thus the social conditions of the practice were not conducive to the further development of forms of practice which were envisaged in the design. The solution to the disaffection with the design on the part of the new leadership resulted in a new design. However, the interview data attest to some evidence of preferences for the original design by the students.

School C4

In School C4, there was little involvement of the original headteacher C4 in the vision, design and construction phases. The data from our interviews showed that the BSF procurement process produced problematic conflicts and discontinuity for the educational considerations of the proposed design. After the design and delivery team was selected for the construction contract, the educational vision was no longer a key priority in the development of the project for the design team. While it was apparent that the school community felt fully involved in the development of the educational vision and conceptual design, the subsequent detailed design and build process actively disengaged the end-users from the design team of the final build.

The acoustic engineer raised major concerns about whether the design of open-plan learning models would function effectively; key decisions were made at the construction phase to cut costs that impacted on the quality and performance of the built design. The aims of the educational vision were significantly compromised and the built design has low-quality acoustic and environmental services specification which are not fit for the purposes of the imagined practice.

The current headteacher C4 has a strong focus on attainment and has been successful in improving standards. However, the school is not managed in a way that aligns with the original educational vision of the design. The timetable does not place same-year groups or subjects in the open learning areas with team teaching. The occupation of the building is now characterised by informal attempts by teachers to change the organisation of space. Teachers have moved old furniture in different ways to try to recreate single classroom spaces in all the open-plan learning areas. This results in physically awkward spaces which are generally regarded by teachers and students as not fit for the purposes for which they are now used. They are particularly problematic acoustically and environmentally in terms of light, ventilation rates, air quality and temperature control.

The second occupation C4 was based on a pedagogic vision which was in stark contradiction to the imagined practice of the original design. Students and teachers are concerned by the significant environmental challenges. Design and practice are in direct conflict. Teachers have commented that using these informally adapted open classrooms disrupts and add tensions to their daily teaching practice:

> I teach with my back to the other open classroom, I absorb anything that is going on there noise-wise. I absorb it, and because it's hitting me first before it's hitting the students in my area, it does throw me when the other class is being disruptive and we can hear everything.
>
> *(Y7 English teacher interview)*

Another Y8 teacher uses a double space as her classroom. The open learning zones were designed to have two or three classes together but now it is just used for one class at a time:

> It only works when I don't have another class teaching in the other open space . . . having no doors and walls is a real problem.
>
> *(Y8 Maths teacher interview)*

Students also feel that the noise within these areas disrupts their learning:

> I don't really like the open-plan because you can hear all this noise and it distracts you all the time. The other class is doing Maths and we are trying to do spellings, it's very confusing.
>
> *(Y7 student interview)*

In this school there was a direct contradiction between design and practice. The voice of the original educational vision for the design disappeared with a change of leadership and there was no representation of the vision in the debates about practice in the school. Prolepsis from design to practice failed. The design was not fit for the purposes for which it was being used and makeshift adaptations failed and in some cases compounded the difficulties experienced by staff and students alike.

Discussion and conclusion

At the outset of this chapter we showed how policy in the early years of the twenty-first century aimed to transform pedagogic practice through the creation of new designs for school buildings. We argued that the notion of proleptic instruction, as inspired by the work of Vygotsky and developed by Cole, can be used to theorise the ways that visions of new practice can be understood in terms of their transformation of the future. In this study we have shown how when the conditions of proleptic instruction are not realised, visions for the future fail to transform practices of the future.

We have shown that if design and practice are in alignment, then design offers a range of possibilities, it invites transformation. If design and practice are in conflict, then practitioners experience significant challenges resulting in dissatisfaction and discomfort.

The structuring of future practice requires a process of learning with the voice of the vision of that transformation. The alignment between the imagined future and the actions of practitioners requires both the artefacts of the future (buildings) and the voices of their proponents. Understood in this way, design alone does not and cannot change practice. It is part of a complex process of transformation through time which in itself must be regarded and enacted as a pedagogic practice.

As we argued in Daniels et al. (2017), it is as if there is a process of resignification at each point of cultural change in successive management regimes. This raises considerable challenges for the kinds of social and cultural transformation that were envisioned in BSF. If the process of prolepsis fails, then the vision fails. If it succeeds, it only does so while the imagination of the practices remains aligned with the vision. If changes in policy or pedagogic predilection of a new school leader drift away from the original vision, then there will be a need for an effective form of adaptation. The original BSF policy documents voiced a concern for adaptation. In practice it appears to be both difficult to enact and often very expensive. This suggests that the notion of sustainability of a school design should place much more emphasis on the possibilities for resignification. Failure to do this can result in more dissatisfaction.

The new BSF designs commissioned by the county's Transformation Team all witnessed social priorities whether through 'heart' spaces, large open spaces at the centre of the schools for social exchange, or the mixed economies of teaching spaces or community spaces for active engagement. The NGT data suggest that if the design is predicated on principles of the importance of social relations, then it is at the level that the proleptic instructional effect as witnessed in the mediation by the artefacts. At the level of the social relations of the instructional practice it would appear to have little effect unless the social conditions of that practice mirror those imagined in the design.

As the National Audit Office (2017, p. 12) argues 'to deliver value for money, the Department must make the best use of the capital funding it has available – by continuing to increase the use of data to inform its funding decisions and by creating places where it can demonstrate that they will have the greatest impact'. This calls for a greater understanding of the relationship between design and educational practices and the impact on experiences of students and teachers.

The methodology that we have developed in the course of this project has shifted the potential gaze of post-occupancy evaluation from a static, and often highly delimited, view of functioning at one moment in time to a much more responsive view of the dynamics of design in-practice over time. An understanding of the ways in which perceptions of students and staff change as they interact within often rapidly changing social situations in which a building, understood as an artefact, is shaped and transformed through time.

Cole's (1996) concept of proleptic instruction in child rearing offers much to the investigation of activities which seek to design interventions in the future of practice. The idea of projecting futures onto practice through design is, of course very appealing, given that it suggests that innovation and development can be brought about through changing spaces. This project suggests we need to understand design in terms of its social origins and the social circumstances of the dynamics of its futures. The social analysis of design is multifaceted. We need to understand the social relations of the design process itself. If the social relations of design are not effective, then the resulting building will not be fit for the purposes of the imagined future. The social characteristics of the practices of design may constrain the social possibilities for the outcomes. We also need to understand the social relations of design in-practice through time.

References

Chapple, M. and Murphy, R. (1996) The Nominal Group Technique: Extending the evaluation of students' teaching and learning experiences, *Assessment and Evaluation in Higher Education*, 21(2): 147–60.

Cole, M. (1996) *Culture in Mind*. Cambridge, MA: Harvard University Press.

Daniels, H. (2001) *Vygotsky and Pedagogy*. London: Routledge.

Daniels, H., Tse, H.M., Stables, A. and Cox, S. (2017) Design as a social practice: The design of new-build schools, *Oxford Review of Education*, 43(6): 767–87.

Daniels, H., Tse, H.M., Ortega, L., Stables, A. and Cox, S. (2018) Changing Schools: A study of primary secondary transfer using Vygotsky and Bernstein.

DfE (2018) *Permanent and Fixed Period Exclusions in England: 2016 to 2017*. London: DfE.

Griffin, P. and Cole, M. (1984) 'Current activity for the future: The zo-ped, in B. Rogoff and J. Wertsch (eds), *Children's Learning in the 'Zone of Proximal Development'*. Jossey Bass: San Francisco, pp. 45–64.

Holland, D. and Lachicotte, W. (2007). Vygotsky, Mead, and the new sociocultural studies of identity, in *The Cambridge Companion to Vygotsky* (pp. 101–35). New York: Cambridge University Press.

Ivic, I. (1989) Profiles of educators: Lev S. Vygotsky (1896–1934), *Prospects*, XIX(3): 427–35.

Kraftl, P. (2006) Building an idea: The material construction of an ideal childhood, *Transactions of the Institute of British Geographers*, 31(4): 488–504.

Lefebvre, H. (1991 [1974]) *The Production of Space*. Cambridge, MA: Blackwell.

Ministry of Education, New Zealand (2011) The New Zealand School Property Strategy 2011–2021. Education.govt.nz. www.education.govt.nz/assets/Documents/Primary-Secondary/Property/SchoolProperty Strategy201121.pdf

National Audit Office (2017) *Capital Funding for Schools*. Report by the Comptroller and Auditor General. London: NAO.

OECD (Organisation for Economic Co-operation and Development) (2010) OECD review of the secondary school modernisation programme in Portugal. CELE Exchange 2010/1. www.oecd.org/education/innovation-education/centreforeffectivelearningenvironmentscele/44708107.pdf

OECD (2017) *The OECD Handbook for Innovative Learning Environments*. Paris: OECD.

Pomeroy, E. (2000). Experiencing Exclusion, *Improving Schools*, 3(3): 10–19.

PricewaterhouseCoopers LLP (2007) *Evaluation of Building Schools for the Future – 1st Annual Report*. London: DCSF.

Reid, K.C. and Stone, A. (1991) Why is cognitive instruction effective? Underlying learning mechanisms, *Remedial and Special Education*, 12(3): 8–19.

Tse, H.M., Learoyd-Smith, S., Stables, A. and Daniels, H. (2014) Continuity and conflict in school design: A case study from Building Schools for the Future, *Intelligent Buildings International*, 7(2–3): 64–82.

Vygotsky, L.S. (1987) *The Collected Works of L.S. Vygotsky. Vol. 1: Problems of General Psychology, Including the Volume Thinking and Speech*, ed. R.W. Rieber and A.S. Carton, trans. N. Minick. New York: Plenum Press.

Credit: Tim Crocker

7
CHANGING SCHOOLS

This chapter is concerned with student experiences of moving between one form of practice and design in a primary school to another form of practice and design in a secondary school. As is the case elsewhere in this book, we deploy the theory of sociogenesis developed by L.S. Vygotsky and the sociology of pedagogy developed by Basil Bernstein, in this case, in order to study the consequences of different trajectories of transfer between different designs and cultures of primary and secondary schools in our sample. Our intention is to contribute to current debates about the effects of new school designs and the enduring concerns raised by difficulties that some students encounter in transitions between schools. Moving between different forms of pedagogic practice that are aligned with design environments appears to have a marked effect on how connected students feel to their schools. It has long been well known that the transition between some homes and school on starting education is more difficult than those where continuity is evident (Douglas, 1964; Melhuish et al., 2008). However, at the outset it is important to note that this literature tends not to discuss the process of transition in terms of the implications of moving between specific cultures or forms of practice. Rather the emphasis is on a more general form of analysis of transition between the primary and the secondary sectors of schooling. Our analysis attempts to identify, analyse and discuss the implications of moving between specific forms of practice in the primary and secondary sectors.

The intention is to open the discussion on how a school building influences experiences of the end-users, how the students are affected by different school pedagogic modalities when changing from one physical space to the other, in our case, from primary school building to newly built secondary schools. Here modality refers to the most common form of practice at a particular site. In any school there will be some variation between classrooms. Even in highly controlled environments subtle nuances creep in to the practices in each classroom. We sought to identify the dominant and most pervasive of the pedagogic actions and arrangements that were witnessed in each school. We build on a short series of previous studies. In Daniels (1995) it was argued that the suggestion that different types of schooling give rise to different types of effect carries with it questions of structural fitness for purpose and that differences in the structure of pedagogic practices constitute differences in contexts which are of semiotic significance. More recently, in Daniels (2010) an analysis of communicative action provided an approach to the consideration of the sequential and contingent development of concepts over time in specific institutions.

Here we were concerned with school designs as explicit and tacit relays of the structure of pedagogic practice as students moved between different structures of pedagogic practice over time. We followed students from the end of Year 6 in their primary schools through their transition into Year 7 of their secondary schools. We were examining the implications of moving from one pedagogic environment to another and one physical/design environment to another. Our concern was the effects of continuities and discontinuities in these trajectories. Thus our focus was on the effects of change in physical and pedagogic factors at the institutional level as experienced at the personal level by students. Chapter 3 provides the details of the data gathering approaches. In this chapter we will examine changes in connectedness scores, interview responses and observations.

We will discuss the general trends witnessed in the quantitative analysis and then proceed to illustrate the qualitative aspects of these transitions as experienced by individuals.

Policy context

The secondary schools sample reported on in this chapter consisted of 11 schools built under the Building Schools for the Future Programme (BSF) or the Academies Programme and seven established older comparator schools. The Academies Programme was announced in March 2002. The intention was to replace an existing failing school or build a new school in an area of sustained low educational performance and expectation. As noted in Chaper 2, BSF, launched in 2004, was the government-sponsored building programme of new secondary schools in England that was in place in the first decade of the twenty-first century.

Aspirations for the outcomes of BSF were couched in terms of collaboration between schools, the development of new forms of infrastructure, new models of school organisation, an enhanced teaching force, new patterns of distributed leadership, personalised approaches to teaching and learning involving significant and novel use of ICT and new forms of central governance (see Hargreaves, 2003).

These new schools were spoken of as 'new cathedrals of learning' (DfES, 2002) in which radical transformations in practices of teaching and learning would take place. The term 'personalisation' was a common feature in many policy documents and although it was linked to a myriad of meanings, generally it became associated with shifts in modes of control over learning with students taking more responsibility for the selection, sequencing and pacing of their work in school. This new form of work required new school designs that afforded the possibility of working in new ways. The personalised approach was to be made feasible through access to new technologies with the availability of a mixed economy of large open-plan and flexible spaces as well as smaller enclosed classrooms. The argument promoted in favour of this significant investment was couched in terms of transformation of learning and teaching along with enhanced participation and community involvement and engagement (BB95, 2002).

The struggles to agree upon what counts as design knowledge and its cultural identity can therefore be perceived as affecting and being affected by a complex system involving economy, production, social significance, consumption, use of objects and so on (Carvalho and Dong, 2006, p. 484).

Theoretical framework

In this instance we draw on the theoretical developments which have influenced and drawn on the work of the British sociologist Basil Bernstein and the Russian social theorist Lev Vygotsky. We do so because we are concerned with interactional, mediated practices and the institutions in which they are enacted. The essence of the developmental model advanced by Vygotsky is a dialectical conception of the relations between the personal and the social. Clearly schooling constitutes a form of collective social activity with specific forms of interpersonal communication. Furthermore within schools and between schools there are differences in the content, structure and function of interpersonal communication. However, a good deal of the post-Vygotskian research conducted in the West has focused exclusively on the effects of interaction at the interpersonal level, with insufficient attention paid to the interrelations between interpersonal and sociocultural levels. Additionally, and perhaps as a consequence of this, schooling is often thought of as a generic activity, as if it were a social institution which is uniform in its psychological effects:

> Vygotsky attached the greatest importance to the content of educational curricula but placed the emphasis on the structural and instrumental aspects of that content . . . In this connection it must be said that Vygotsky did not take these fruitful ideas far enough. In this approach it is quite possible to regard the school itself as a 'message', that is, a fundamental factor of education, because, as an institution and quite apart from the content of its teaching, it implies a certain *structuring of time and space* and *is based on a system of social relations* (between pupils and teacher, between the pupils themselves, between the school and it surroundings, and so on).
>
> *(Ivic, 1989, p. 434)*

Artefacts are products of human history that serve to bring together the cultural historical, the institutional and the personal levels. Cole (1998) argues that artefacts not only 'change our conditions of existence, but also act on us and cause change in our mental condition' (Luria, 1928, p. 493). Following Vygotsky (1987), he argued that interpersonal processes are transformed into intrapersonal processes as development progresses and that there is a mutual shaping of person and context:

> These units of analysis therefore integrate the micro-social socio contexts of interaction with the broader social, cultural and historical contexts that encompass them.
>
> *(Tudge and Winterhoff, 1993, p. 67)*

Abreu and Elbers (2005, p. 4) further argue that in order to understand social mediation it is necessary to take into account ways in which the practices of a community are structured by their institutional context. The mutual shaping of person and place is recognised by Burke, who suggests that the 'vision of school as a transformed space for learning . . . could not exist separately from a transformation in the view of the child as artist of their own learning and builder of their own worlds' (Burke, 2010, p. 79). However, there are very few examples of a Vygotskian analysis of school architecture as a structuring resource (Flygt, 2009). Accordingly we investigated the ways in which the design of space within schools mediates and shapes practices of teaching and learning.

As noted in Chapter 2, Moos (1979) argued that the learning environment is best understood as resulting from a complex interaction of social, cultural, organisational and physical factors. Benito (2003) directed attention to the meanings of school design and the cultural function that is assigned to schools. From this perspective school architecture should be open to a form of analysis which takes account of educational discourses and practices, and actors' social norms.

Vygotskian theorists argue that individual agency has been significantly under acknowledged in Bernstein's sociology of pedagogy (e.g. Wertsch, 1998). Vygotsky's work provides a compatible account which places emphasis on individual agency through its attention to the notion of mediation. Sociologists complain that post-Vygotskian psychology is particularly weak in addressing relations between local, interactional contexts of 'activity' and 'mediation', where meaning is produced and wider structures of the division of labour and institutional organisation act to specify social positions and their differentiated orientation to activities and 'cultural artefacts' (e.g. Fitz, 2007). Vygotsky offers little by way of an approach, such as that of Bernstein, to the analysis and description that enables researchers to relate macro-institutional forms to micro-interactional levels.

Bernstein provides a semiotic account of cultural transmission which is avowedly sociological in its conception. In turn the psychological account that has developed in the wake of Vygotsky's writing offers a model of aspects of the social formation of mind which is underdeveloped in Bernstein's work.

There was a need to refine a language of description that would allow our research to 'see' institutions as they did their tacit psychological work through the discursive practices that they shaped. A way of describing what were essentially the pedagogic modalities of the settings in which we were intervening was required. That is, the most likely forms of institutional practice that would be sustained in those settings. These mediate social relations and shape both thinking and feeling: the 'what' and 'how' as well the 'why' and 'where to' of practice. We were concerned with the ways in which wider social structures impact on the interactions between the participants.

We also recognised the importance of developing an approach to the analysis and description of our research sites that could be used to monitor changes that took place over the course of our study. These understandings formed the background to the development of an account of institutional structures as cultural historical products (artefacts), which play a part in implicit (Wertsch, 2007) or invisible (Bernstein, 2000) mediation.

In his analysis and thus his descriptions of schools, Bernstein (1977) focuses upon two levels in his account of cultural transmission; a structural level and an interactional level. The structural level is analysed in terms of the social division of labour it creates and the interactional with the form of social relation it creates. The social division of labour is analysed in terms of strength of the boundary of its divisions, that is, with respect to the degree of specialisation. Thus the key concept at the structural level is the concept of boundary, and structures are distinguished in terms of their category relations. The interactional level emerges as the regulation of the transmission/acquisition relation between teacher and taught, that is, the interactional level comes to refer to the pedagogic context and the social relations of the classroom or its equivalent. The curriculum may then be analysed as an example of a social division of labour and pedagogic practice as its constituent social relations through which the specialisation of that social division (subjects, units of the curriculum) are transmitted and expected to be acquired.

Power is spoken of in terms of classification which is manifested in category relations. At the macro level, classification generates categories of agents and discourses: the categories or insulations are instantiations of power. At the micro level, classification is about the organisational or structural aspects of pedagogic practice. Classification is about relations between, and the degree of maintenance between categories, and these include the boundaries between agents, spaces and discourses. Control may be spoken of in terms of framing which is manifested in pedagogic communication. The specialised form of communication whereby differential transmission and acquisition is effected is the pedagogic discourse (Bernstein, 1990, p. 182). Framing, therefore, refers to relations within (within boundaries). As Hoadley (2006) argues, framing, in a sense, supports classification: it produces 'the animation of the power grid' (Hasan, 2002), but also opens up the potential for the change of boundaries, the contesting of power relations. It is through interaction (framing) that boundaries between discourses, spaces and subjects are defined, maintained and changed. In our study we have taken this work as a point of departure in the development of a model of description.

Daniels (1989) utilised the distinction made by Bernstein (1977) between instructional and regulative discourse. The former refers to the transmission of skills and their relation to each other, and the latter refers to the principles of social order, relation and identity. Pedagogic discourse is defined as the rule which embeds a discourse of competence (the instructional, including specific skills) into a regulatory discourse (regulatory of character, conduct and manner, and of theories of pedagogy). The distribution of power and principles of control differently specialise structural features and their pedagogic communicative relays. The instructional is embedded in the regulative and this means that the hierarchical relation between transmitter and acquirer regulates the selection, sequencing, pace and evaluative criteria of the instructional knowledge.

The central generative construct which underpins these aspects of schooling is the dominant theory of instruction that is explicitly or tacitly adhered to by designers and practitioners. As Bernstein states:

> The theory of instruction is a crucial recontextualised discourse as it regulates the orderings of pedagogic practice, constructs the model of the pedagogic subject (the acquirer), the model of the transmitter, the model of the pedagogic context *and* the model of communicative pedagogic competence.
>
> *(Bernstein, 1985, p. 14)*

Where the theory of instruction gives rise to a strong classification and strong framing of the pedagogic practice, it is expected that there will be a separation of discourses (school subjects), an emphasis upon acquisition of specialised skills, the teacher will be dominant in the formulation of intended learning and the pupils are constrained by the teacher's practice. The relatively strong control on the pupils' learning, itself, acts as a means of maintaining order in the context in which the learning takes place. The form of the instructional discourse contains regulative functions. With strong classification and framing the social relations between teachers and pupils will be more asymmetrical, that is, more clearly hierarchical. In this instance the regulative discourse and its practice is more explicit and distinguishable from the instructional discourse. Where the theory of instruction gives rise to a weak classification and weak framing of the practice, then children will be encouraged to be active in the classroom, to undertake enquiries and perhaps to work in groups at their own pace. Here the relations between teacher and pupils will have the appearance of being more symmetrical. In these circumstances it is difficult to separate instructional discourse from regulative discourse as these are mutually embedded. Relatively little has been done to apply Bernstein's work to study the effects of school design on pedagogic practices/spatial affordances for pedagogic practices. Although Bernstein does make a passing reference to the influence of Piagetian theory and pedagogic practice on the design of primary schools in the 1960s:

> The fit between Piaget (E) and the ideological emphasis in primary education, with respect to the definition of the child as a pedagogic subject, the model of the pedagogic context, its interactive format, legitimate texts and their mode of creation and evaluation, was mutually reinforcing. The architecture of the new schools with their weak boundaries and transparencies resonated with the new model. Of importance, the 1960s were a period of expansion of all resources in education partly as a response to population pressure and state welfare policy. Thus the concept of competence legitimated and constructed the de-contextualised but active, creative child, abstracted from gender, class, race, region, apparently the imaginative author of his or her texts under the aegis of internal motivation and peer group activities.
>
> *(Bernstein, 1993, p. xii)*

Our concern has been to progress the development of an account of the mediated effects of schools design on groups and individuals as they move through time and between spaces. To do this we draw on both Vygotsky and Bernstein and in so doing hope to contribute to a more refined theoretical argument.

Hoadley (2006) provides an example of the application of Bernstein's work to the study of space in traditional schools in South Africa. She refers to teacher – learner spaces (strength of demarcation between spaces used by teachers and learners) and space for learning (strength of boundary between space, internal and external, to the classroom and learning). In this we are concerned with overall school design and this involves a much broader conception of space and innovation in school design. We were also concerned with the progressive recontextualisation of the design through subsequent occupations of the school as new leaders (headteachers) were appointed.

Vygotsky, Bernstein and Design

From Vygotsky (1987) we argue that engaging in the pedagogic discourses (thought of as tools) and practices of each school transforms the activity of schooling and gives rise to specific orientations to meaning. These are the tools which mediate thinking and feeling and are in turn shaped and transformed through their use in the activity of schooling. In this way processes of co-creation of individual/psychological and cultural/historical factors become interwoven. From Bernstein (2000) we developed an account of the regulation of these discourses and practices as institutional modalities.

Where the theory of instruction gives rise to a strong classification and strong framing of the pedagogic practice, the spaces used for instruction would be expected to be strongly demarcated. Single-cell classrooms designed for single classes of students would be expected. The relatively strong control on the pupils' learning, itself, acts as a means of maintaining order in the context in which the learning takes place. The form of the instructional discourse contains regulative functions. With strong classification and framing the social relations between teachers and pupils will be more asymmetrical, that is, more clearly hierarchical. As in Hoadley's (2006) study, there would be an expectation that the teacher would occupy space at the front of such classrooms. In this instance the regulative discourse and its practice is more explicit and distinguishable from the instructional discourse. Where the theory of instruction gives rise to a weak classification and weak framing of the practice, then children will be encouraged to be active in the classroom, to undertake enquiries and perhaps to work in groups at their own pace. In this version of a personalised approach, curriculum subjects may be abandoned in favour of themes to be explored through project-based enquiry. A mixed economy of spaces with large open areas and smaller breakout spaces for small group or individual study would facilitate this form of pedagogic practice. Here the relations between teacher and pupils will have the appearance of being more symmetrical. Teachers would be unlikely to retain 'ownership' of particular spaces. In these circumstances it is difficult to separate instructional discourse from regulative discourse as these are mutually embedded.

Methodology

In the *Design Matters?* project we used Bernstein's work, inter alia, to develop an approach to the analysis and description of the schools as modalities of institutional practice which was used in the subsequent analysis of data concerning experiences of occupation as the designs were transformed under different theories of instruction as promoted by successive headteachers. In so doing it examines the relationships between the structuring of space in a building, the structuring of social relations and practices and the psychological consequences for occupants of the building.

Student-level data

In this chapter we only report the data concerning practice as enacted at the times at which the students made their transition from primary school. One of the several approaches to gathering user perceptions was to survey student responses to a school connectedness survey. 'School connectedness' is a concept that has been used in a variety of ways as an attempt to identify the psychological 'fit' of students to the school environment, encompassing elements such as health, security, social relations and self-esteem.

The device we employed was modified slightly from that developed by Goodenow (1993). Goodenow developed a measure of youth connectedness to school, showing it to have high internal validity, with a Cronbach's Alpha score of 0.88. On this measure, the more nearly the score reaches 1.0, the more the items in the scale can be trusted to form a consistent measure of the construct under investigation. Goodenow devised the scale for use with 12 to 18 year olds, whereas the *Design Matters?* team used it with 11 to 13 year olds. A trial resulted in a slight reduction of the number of items, where we felt there was some degree of confusion among students about what an item meant. Our scale thus comprised 11 items. We also supplemented the five-point answering boxes (from 'Not at all True' to 'Completely True') with 'smiley' emojis showing a range of emotions connected with the relevant response. In other ways, the scale remained true to Goodenow's original. The questionnaire is reproduced in Appendix.

This was our only strictly quantitative measure (though we could also quantify findings from other sources), and so our major source of statistically robust evidence about students' feelings prior to, on entry to, and at the end of their first year of secondary school.

School- and student-level data were gathered during the final term of Year 6 (time 1, n = 452) in the primary schools and the first (time 2, n = 498) and third terms of Year 7 (time 3 n = 415) in the secondary schools. The primary schools acted as feeder schools for both new-build and established schools in each locality. The data were collated in such a way that, where possible, the student-level data could be analysed as a collection of individual trajectories from school to school.

School-level data

The general model of description of the institutional modality of the schools was developed from a Bernsteinian perspective under the headings:

- School Design
- Pedagogic Practice as Enacted in the Design
- External Relations as Enacted in the Design.

School Design was understood in terms of an 'instructional element' in which the classification of space was modelled as a design and subsequently remodelled in practice. This classification of space was associated with a measure of framing. The regulative aspect of this discourse was concerned the explicit (positional) or implicit (personal) regulation of the general social order through the building.

The Pedagogic Practice as Enacted in the Design was also analysed in terms of classification (the organisation of teaching for curriculum subjects and grouping of students (numbers of students in a teaching unit for example, traditional class of 30+ or multi unit of 90 or 120)) and framing of the practice (the extent to which the practice was personalised in terms of curriculum selection, sequencing and criteria of evaluation). The regulative aspect of this part referred to the extent to which the social order, identity and relation in the teaching space was the object of implicit or explicit control.

The External Relations as Enacted in the Design aspect of the model referred to the degree of insulation between the school and community (classification) and the degree of control over the internal/external relation.

Following Bernstein, in the coding instrument, the high-level concepts of classification and framing were translated into a coding scheme to read the data. The indicators, or theoretical constructs, named empirical instances of particular abstract concepts. The coding was performed using a four-level scale where ++ represents strongest and − represents weakest. The scale was ++, +, -, − and applied to values of classification (C) and framing (F). Clearly there were no absolute measures which applied. The purpose was to use descriptions which would demarcate the schools from one another and draw attention to important characteristics.

The coding of each school in terms of specific classification (strength of category relation) and framing (social relation) values was based upon observation and interview data. For each school we developed descriptions based on the model which incorporate the data gathered from the wide variety of sources. This allowed us to consider the extent to which the original design was witnessed in practice at each occupation of the school by consecutive headteachers. We were also able to reflect on the relationship between the accounts of practice that were gathered and observations of practice that we made.

Each school was visited at least six times and interviews were conducted with headteachers, teachers and students. In secondary schools facilities managers and parents were also interviewed both individually and in focus groups. Teaching areas were observed as was the use of space at break and lunch time. A tour of the school was conducted with the architects involved in the design of the building. Table 7.1 provides examples of elements of the coding frame.

Two broad groupings of data emerged from this extended qualitative analysis. Each grouping was quite broad nevertheless there was a clear distinction between the two modalities. We were particularly interested in settings in which the regulation of the practice envisioned in the school design was over ridden or subverted by the pedagogic practice as enacted.

In Modality A, the instructional element of School Design the classification of space was seen to be strongly classified with strong framing. The regulative aspect of this discourse was concerned the explicit (positional) regulation of the general social order through the building.

In Modality A, Pedagogic Practice as Enacted in the Design the classification of space was strong and the framing of the practice in terms of curriculum selection, sequencing and criteria of evaluation was

TABLE 7.1 Examples of elements of the coding frame for classification of spaces as designed and framing over use of space as envisaged in the design

Coding of organisation of space	C-	C-	C+	C++
Indicators	Large open learning plan spaces for multiple groupings (60–120 students) with breakout spaces for individual and small group work	A small number of large open spaces and some enclosed classrooms. Designed for flexible group sizes	Enclosed classrooms with limited availability of additional space outside the classroom for social and/or pedagogic purposes. Designed for single class groups	All enclosed classrooms designed for sole use of single groups of approx. 30 students
Coding of control envisaged in use of space in school design	F-	F-	F+	F++
Indicators	Learners choose how use and move through the spaces	Teachers influence learners decisions about use of space and movement	Teachers discuss but direct eventual use of space and movement	Teachers direct use of space and movement

also strong. The regulative aspect was based on explicit control of social order, identity and relation in the teaching space.

In Modality A, the External Relations as Enacted in the Design revealed strong insulation between the school and community and strong control over the internal/external relation.

In Modality B the School Design was weakly classified with a mixed economy of very large open spaces and smaller enclosed spaces as promoted under the BSF guidance (BB95, 2002). It was also weakly framed in line with the personalisation argument in which students were supposed to exercise control of the selection, sequencing and pacing of their studies within a thematic approach to the curriculum. However, we found several examples of this modality which were subsequently adapted as the modality of pedagogic practice changed through time and with subsequent changes of leadership. Details of these adaptations are the focus of another publication (Daniels et al., 2017). The regulative aspect of Modality B was concerned with the implicit (personal) regulation of the general social order through the building.

The Pedagogic Practice as Enacted in the Design of Modality was aligned with features outlined in the Modality B design. However, there were several examples of practices of attempts at strengthening classification through the building or improvisation of barriers or walls in the open spaces in order to enact a modality a practice. In these cases the modality of the design was clearly not aligned with the modality of the practice. In the regulative aspect of Modality B the social order, identity and relation in the teaching space was the object of implicit control. Again this aspect was subject to adaptation over time.

The External Relations as Enacted in the Design of Modality B involved weak insulation between the school and community and the lower control over the internal/external relation as envisioned the BSF agenda (BB95, 2002). In one of the schools we observed a change from a situation where parents were encouraged into the school at all times and offered free breakfast in order to encourage communication with staff and students to the introduction of a rule that parents were not allowed on the school site.

We have selected the following quotes from two headteachers one of each of Modality A and B schools to provide more of a flavour of the differences. We are not suggesting that one modality is better than other merely that they are different.

The quotes relating to use of space reveal stark differences on the values of classification. For the Modality A headteacher the primary unit for teaching was the single classroom:

> I think the idea behind it is that the children could be flexible and could come in and out, you know, if they want to use a computer or do some group work, those sorts of things, and that doesn't work for our children. The theory behind it and the practice of our children. What they are used for now is to whole classes that . . . well most of the time it's whole classes or at least half a class with a member of staff. And it really does work.
>
> *(Modality A headteacher)*

Whereas the Modality B teacher argued in favour of a mixed economy of space that could be used flexibly:

> I think it increases the quality of teaching because it's so transparent, and the lessons improve and I would say behaviour improves to an extent rather than being shut away in a classroom. But I think we've got a healthy combination here because we've got some classrooms and we've got some open areas and what you will see here is, we haven't gone for kind of whole scale open learning for everybody, it's a combination . . . And there are occasions where a teacher and an assistant will teach a big group and then they will split off into smaller groups, and I think it's perfect for that

kind of thing really. . . . So I think you need a healthy combination. You don't necessarily have to go for a whole scale, whole curriculum open learning, and you couldn't here anyway because there are lots of classrooms, but I like the combination that we've got.

(Modality B headteacher)

There were also quite different accounts of children, their behaviour and the school's responses to them. The Modality A headteacher presented an account of individuals who required a highly controlled environment:

But a lot of them at times don't respect the building, they don't respect what's been done for them, they see it as prison, they see the uniform as too posh, they will trash the building, they will throw litter on the floor. It's taken me a long time to get the children to sit at a table and eat and to have a tray. I still haven't cracked the fact they will all have a tray and they will all pick up it up. A very simple request but they will do things like they will open a sandwich and they'll take the tomato out that they don't like and they will throw it on the floor, they won't put it on the tray or even the table. So it's those basic things that you take for granted that we're still trying to do.

(Modality A headteacher)

Whereas the Modality B headteacher, while not eschewing the need for control, understood the social relations of schooling in a more communitarian manner. This reveals differences in the values of framing that underpin the practices of the two schools:

When we moved in the student behaviour changed literally overnight, the way the staff interacted with the students changed, the community, the parents, how the staff felt coming into the school every day. I could stand on the balcony and see the majority of the school changing lessons. There was nowhere to hide for a student and it was all so visible. Now that can sound punitive in terms of monitoring behaviour but it's also a very friendly environment so you could say hello to people and wave to people from the balcony and usher them along, you know, it's just a fantastic school design and of course there are things we could tweak without a doubt but overall I just think with the heart space there it is exactly what we wanted it to be about, about a community, families coming together and that heart space, the fact that we've developed the vocabulary for our school, to personalise it, it's a lovely school to go in.

(Modality B headteacher)

Accounts of local communities, families and relations of control reveal similar differences. In Modality A the headteacher stated:

They are children who have come from families, third generation of don't like school, there's still a perception when you talk about family, there's still sharp intake of breath. There's still a lot of children who don't value education and the parents don't value education. And we still are reactive on a day-to-day basis. Sometimes you will walk out there and it's bliss, but a lot of the time you will have anything up to 20 children who refuse to go to lessons, that don't want to do what they're told, that don't want to follow instructions. More than just the odd naughty child. You've then got the complexities of special needs. We have a lot of damaged children.

(Modality A headteacher)

Whereas in the Modality B school the emphasis on community was revealed with again weaker values of framing:

> I would say that our role is to help the community, you know, come on the journey with us, to have higher aspirations for the young people. That's what our role is and I don't . . . I wouldn't really want that kind of negative mindset or standpoint. And I don't think many of our staff have that either . . . we've got plenty of areas where we've had older people working with our students or being taught computer skills by our students, and then those older people from the community coming in to do things like paired reading with students, so that's happening now.
>
> *(Modality B headteacher)*

Using the codings we derived from multiple sources of interviews repeated over a year with parents, teachers, headteachers and support staff, observations of lessons and social times inside and outside the schools, we were able to analyse the connectedness data in terms of transition between modalities over time and the implications of settings in which modalities of design were not aligned with modalities of practice.

Findings

We used the modified scale (see Appendix) of 11 Likert scale items also with high internal consistency (Cronbach's Alpha = .85). Item 9 of the modified scale (It is easy to find my way around this school) was not included in the final school connectedness scale as the exploratory factor analysis showed that it was not part of the same component as the other items, and was therefore not considered in the analysis below. A final factor analysis confirmed uni-dimensionality of the scale we used. High scores are more frequent than low scores (i.e. the distributions are negatively skewed). The analysis reported here is of data gathered in the second of two cohorts. The sample size of the first cohort proved to be too small to allow for satisfactory analysis but showed similar trends to those reported here.

Overall, school connectedness by measurement occasion

Overall, as shown in Figure 7.1, school connectedness scores significantly decreased over time. By comparing scores between consecutive time points using independent t-tests, we can see that, on average, student showed higher scores in Time 1 ($M = 4.12$, $SE = 0.03$) than in Time 2 ($M = 3.90$, $SE = 0.03$), $t(424) = 6.42$, $p < .001$. Scores in Time 2, in turn, were significantly higher than scores in Time 3 ($M = 3.54$, $SE = 0.03$), $t(406) = 14.75$, $p < .001$. In terms of effect sizes, the overall decrease in school connectedness scores from Time 1 to Time 2 represents a medium effect ($r = .30$) while the decrease from Time 2 to Time 3 represents a large effect ($r = .59$). This echoes the general trend observed in the literature; for example, Niehaus' data was collected on 330 Year 7 students, aged 11/12, each filling out a questionnaire at the beginning of each academic term (Niehaus et al., 2012). When evidence was collated it was found that there was on average a 7.7 per cent decline in school connectedness, this was consistent across races and genders (Niehaus et al., 2012)

School-level analysis

When examined at the school level, schools' mean connectedness trajectories show considerable variation. This observation, represented graphically in Figure 7.2, is supported by the results obtained after performing a mixed analysis of variance. There was a significant main effect of time, $F(2, 784) = 151.26$,

150 Changing schools

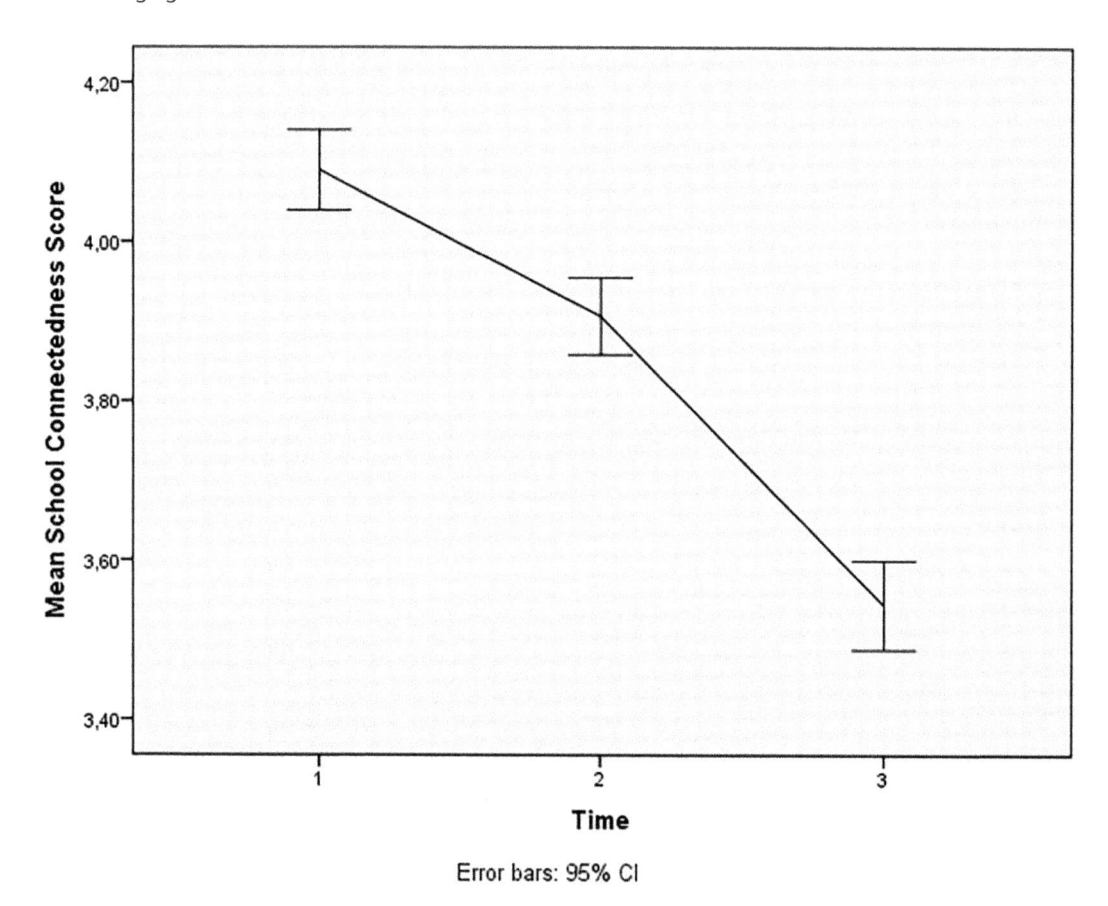

[Figure: Line graph showing Mean School Connectedness Score decreasing across Time 1 (~4.09), Time 2 (~3.90), and Time 3 (~3.54), with 95% CI error bars.]

Error bars: 95% CI

FIGURE 7.1 Overall, connectedness by measurement occasion

$p < 0.001$, a significant main effect of school $F(14, 392) = 16.83, p < 0.001$ and a significant interaction effect between time and school, $F(28, 784) = 59.33, p < 0.001$. This indicates that the change in school connectedness scores over time differed between schools.

New/Established Comparator Schools Analysis

We then compared the school connectedness scores of new and established comparator schools at each of the time points measured using independent t-tests and found that only in Time 2 school connectedness scores of new schools ($M = 3.95, SE = 0.03$) and established comparator schools ($M = 3.58, SE = 0.07$) differed significantly $t(496) = -4.75, p < .001$. In terms of effect size, this difference represents a small to medium effect ($r = .21$). Neither in Time 1 nor in Time 3 were the school connectedness scores of new and established comparator schools significantly different. These comparisons are presented in Figure 7.3.

Modality Analysis

In order to assess whether the different modality groups show significantly different school connectedness scores at each of the time points, we performed one-way analysis of variance (ANOVAs). We found

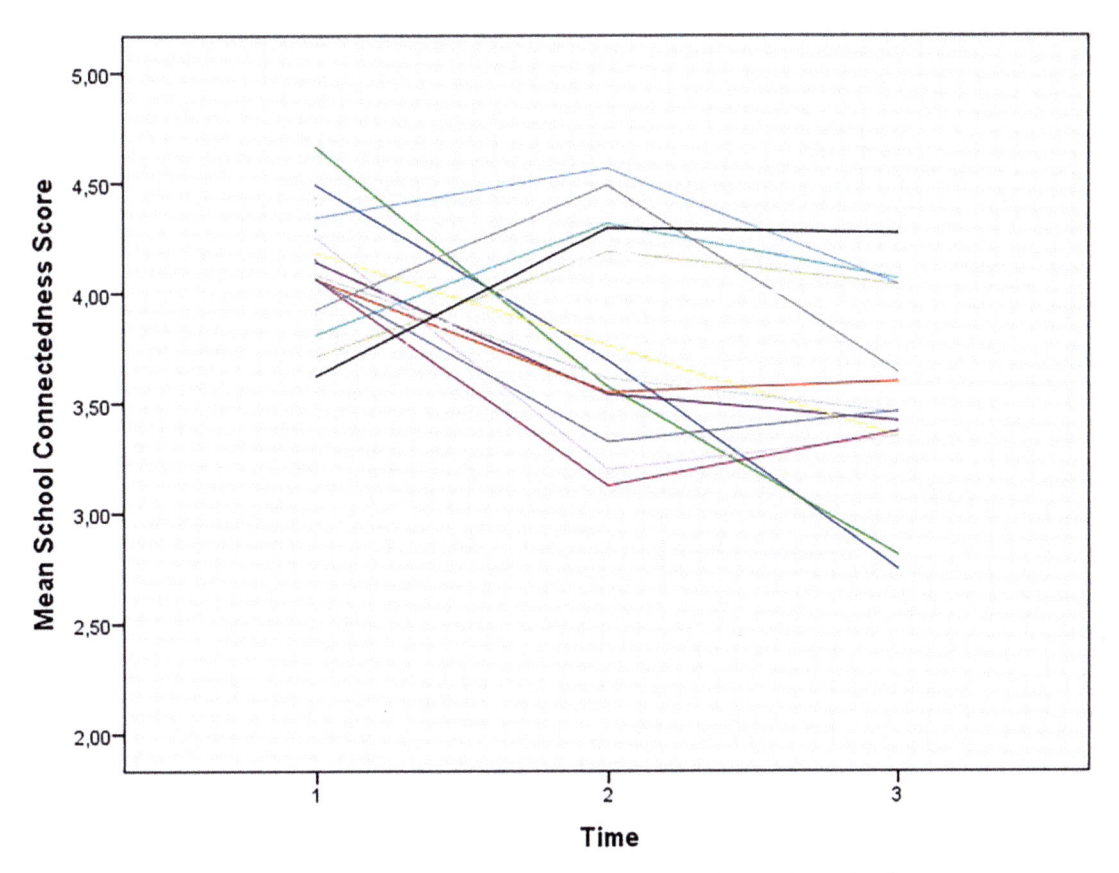

FIGURE 7.2 School connectedness scores by measurement occasion and secondary school

a significant effect of modality group on school connectedness levels at Time 1 ($F(3,374) = 48.02, p < .001, r = .62$), Time 2 ($F(3,358) = 205.57, p < .001, r = 1.31$) and Time 3 ($F(3,349) = 129.58, p < .001, r = 1.06$). The effect sizes of these differences are large.

To assess the extent to which the mean school connectedness score trajectories over time differed between school modalities, we conducted a mixed analysis of variance (ANOVA). We found a significant main effect of modality ($F(3,342) = 100.00, p < .001$), a significant main effect of time ($F(2,684) = 211.86, p < .001$), and a significant interaction effect between modality and time ($F(6,684) = 112.96, p < .001$), which means that the different modality groups progressed differently over time. Looking at Figure 7.4, it is possible to observe that modalities A – A and B – A share similar school connectedness trajectories, with scores steadily decreasing from Time 1 to T 3. Modality B – B starts at a similar school connectedness level but increases in Time 2 and declines somewhat in Time 3. Modality A – -B, in turn, starts lower than the rest of modality groups but then increases significantly in Time 2 and scores remain comparatively high in Time 3.

Design/Practice Alignment Analysis

Using independent t-tests, we compared school connectedness scores of schools where design is in line with practice and schools where design is in conflict with practice. Overall, school connectedness scores

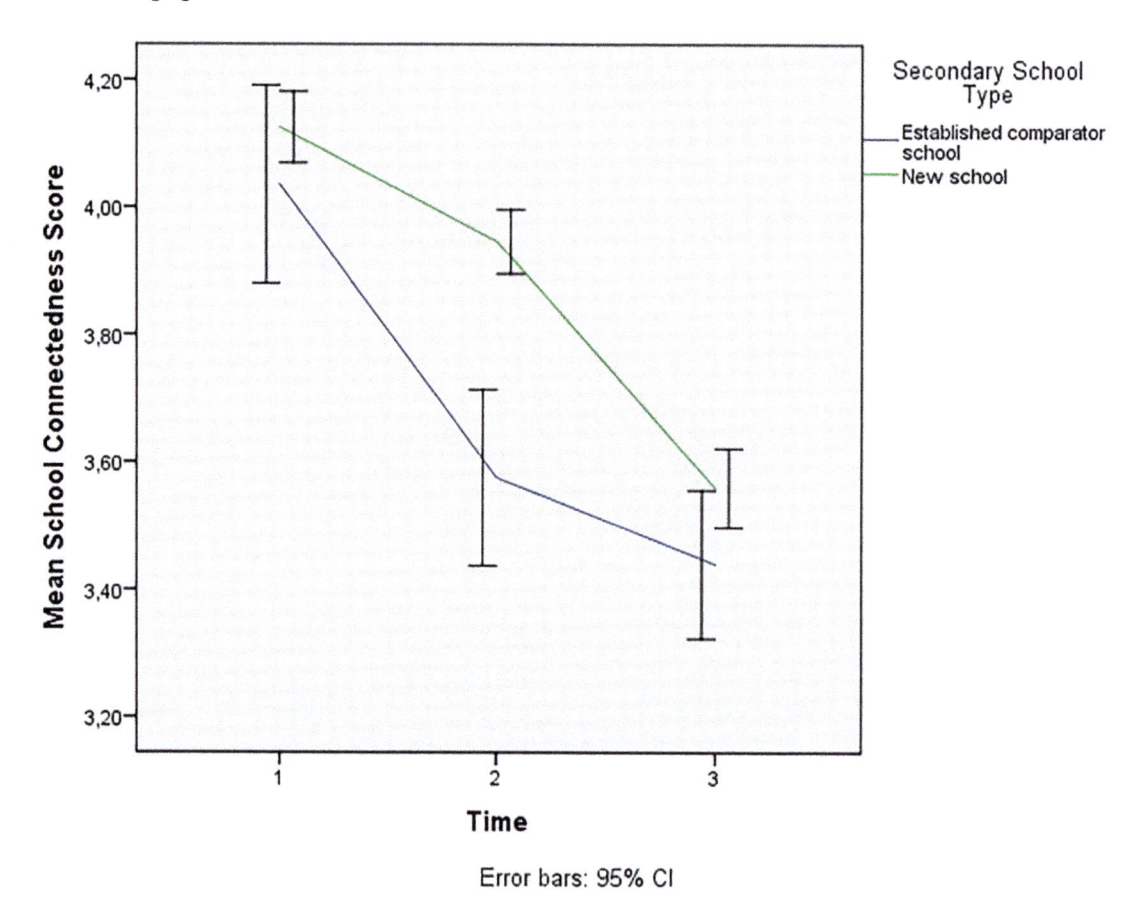

FIGURE 7.3 School connectedness scores by measurement occasion and type of secondary school

of schools where design is in line with practice (M = 4.46, SE = 0.04) and schools where design is in conflict with practice (M = 3.85, SE = 0.03) differed significantly in Time 2, $t(471)$ = -9.99, $p < .001$, r = .42, with schools where design is in line with practice showing higher scores. A similar trend was found in school connectedness scores in Time 3, as schools where design is in line with practice (M = 4.09, SE = 0.03) showed significantly higher scores than schools where design is in conflict with practice (M = 3.43, SE = 0.03), $t(388)$ = -9.55, $p < .001$, r = .42). In terms of effect sizes, this difference represents medium to large effects, as shown in Figure 7.5

To look at differences at the item level between schools where design is in alignment with practice and schools where design is in conflict with practice, we used Mann-Whitney U tests and found no statistically significant differences at Time 1 for any of the items.

Group experiences of transition

Before presenting accounts of individual experiences of transition, we will first provide an over of the aggregated data for the NGT tasks.

During the first task, primary school students were asked to write down what they thought would be different in their secondary school. Their answers were then written on a flipchart and they were asked

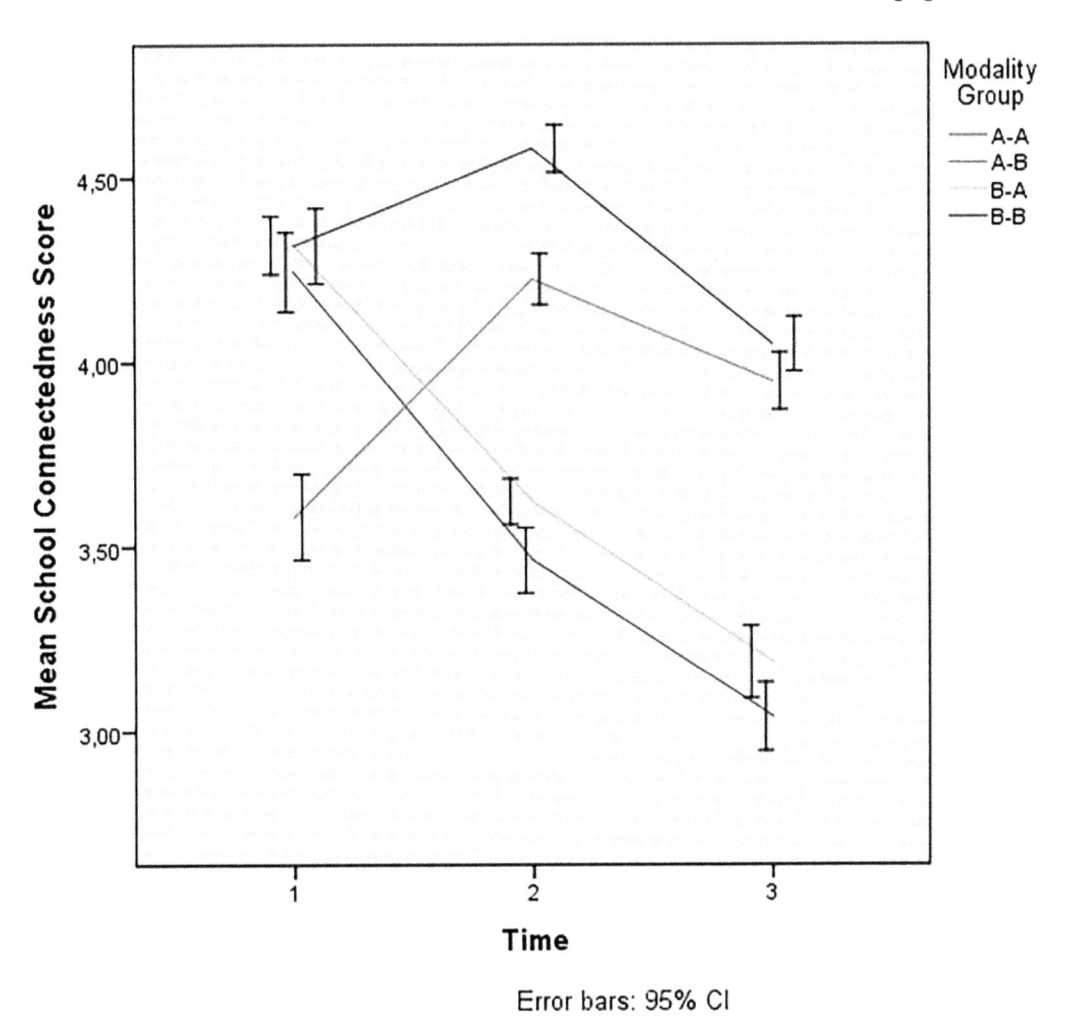

FIGURE 7.4 School connectedness scores by measurement occasion and school modality group

to rank the items according to their importance and come to a consensus. Finally, they were also asked to rank their own original answers.

Overall the results show that the students mentioned changes related to social relations the most. Other themes that emerged in the analysis include schoolwork, logistics, place and personal well-being. Pupils who were moving to new-build schools for their secondary education mentioned social relations more compared to pupils moving to the traditional schools, however, the difference between was a small one (35 per cent for new build, 31 per cent traditional). When the answers were grouped according to school modalities, it was found that pupils moving from Type B primaries to Type A secondaries made the least references to social relations, whereas pupils going from Type A to Type B made the most.

Pupils moving to Type A schools mentioned social relations the least, which might be due to them having experienced highly controlled social environments in their primary schooling. Modality 'B to A' transition students had prioritised changes related to logistics, that is, lockers, uniform and travel more than other students. On the other hand, pupils moving to Type B schools mentioned the social changes the

154 Changing schools

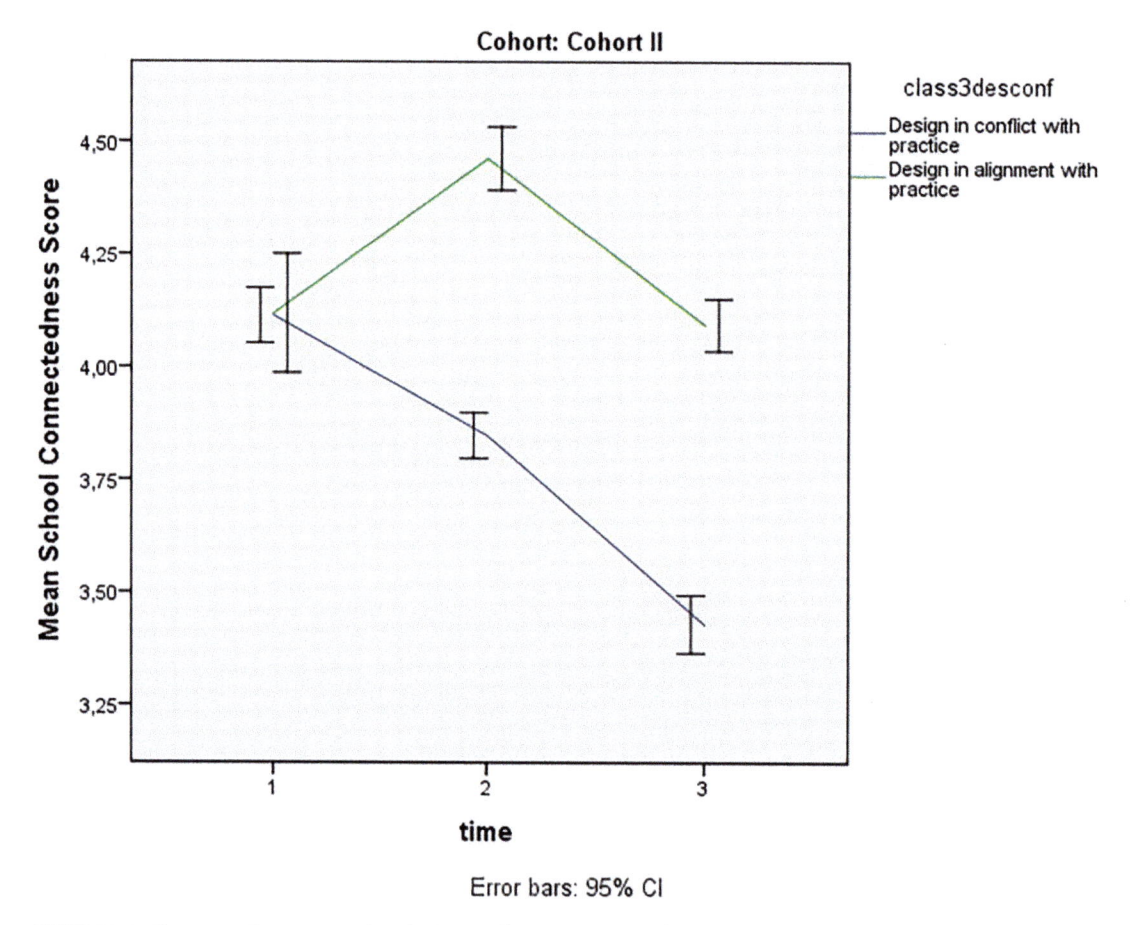

FIGURE 7.5 Connectedness scores in relation to design/practice alignment/conflict

most. Since 'A to B' pupils would be moving to more weakly classified and framed schools from highly framed environments, they might have thought reduced restrictions would have a significant impact on their social life.

Upon starting secondary education, students were asked to write an essay on their first week at their new school. The question given to students was deliberately very open. It was anticipated that references in the texts would reveal matters of priority and concern. Social issues emerge as a significant theme again in these essays, references to student relations being the most frequently mentioned in four out of six schools in the sample. Students cite having friends as an essential factor in helping them settle. There are references about older children; sometimes these are positive in case of mentors who have been friendly and helpful and sometimes negative as they came across as 'inconsiderate', or 'loud'. Overall, 50 per cent of the references about other students were positive, with 29.6 per cent neutral, 18.25 per cent negative and 2.1 per cent combined. Students also made references to their interactions with teachers; 58.4 per cent of these references were positive (32.7 per cent neutral, 6.65 per cent negative, 2.22 per cent combined). Students who had moved to traditional secondary buildings made fewer references to social relations when compared to students who had moved to new buildings (32 per cent traditional, 41 per cent per cent new build).

After the first month of secondary school, the pupils were asked the question 'what was different about your new school?' According to students, changes in their social life were most significant. This was attributed to the design and size of their new schools and the secondary school organisation. Their secondary schools have more spaces to be social both inside the building and outside. Similarly, the movement between lessons was seen to make it easier to catch up with friends.

Students who came from a Type A primary to a Type B secondary made more references to social relations. They referenced social relations in 47.30 per cent of their answers while students who moved from a Type B primary to Type A secondary schools mentioned aspects of physical space more. They mentioned social relations in 22 per cent of their answers, with space references constituting 37 per cent of their answers. During the NGT 2 discussion, students who moved to Type A secondary schools from Type B primaries expressed their frustration with strict dress codes and rules in addition to problems they experienced with older students.

When asked to elaborate on places and design features of their secondary schools that were important to them, pupils in traditional and new buildings differed in which spaces they value. Pupils in traditional buildings prefer academic lesson spaces (28 per cent social spaces, 56 per cent academic spaces), while pupils in new builds valued social and academic spaces equally (44 per cent social spaces, 45 per cent academic spaces).

Overall, Multi Use Games Areas (MUGA) and the heart space were most frequently mentioned spaces for socialising in new buildings. Female students noted that boys have more spaces to be with their friends, as boys tend to use sports facilities during break and lunch. The heart space seems to be important for both male and female students. Students use both outside and inside spaces to socialise. They particularly like open spaces where they can be active with their friends. Similarly, they also like learning places where they can be active.

Individual experiences of transition

In order to gain a sense of how the various data sources contributed to an understanding of individual children's evolving experiences, a number of mini-case studies were extracted from the data. These collations of data bring together the individual's responses to the essay task and the various focus group tasks: the repeated NGT, the school connectedness questionnaires and the map tasks as well as contemporary Ofsted reports. We will not reference these reports as they would identify the schools which have been assured of anonymity in publications. Below, for example, is Jane's story, illustrative (though not necessarily typical) of a child who moved from a highly controlled Modality A primary school into a Modality B secondary environment.

Jane: modality A to modality B (student's name has been changed)

Jane attended CP1 Junior School, which had 483 students at the time of writing. The school's intake is ethnically diverse (9.1 per cent of pupils have a first language that is not English), but the majority of pupils are of white British heritage. At the time of the study the proportion of pupils supported by a school action plus or with a statement of special educational needs is higher than the national average (9.1 per cent compared to 7.7 per cent). However, the proportion of pupils eligible for pupil premium was below average (20.4 per cent compared to 26.6 per cent).

When Jane started at Riverview Junior School, it held a rating of 'satisfactory' from Ofsted. When Jane was in Year 4, the school was inspected again and the rating was raised to 'good', with a number of 'outstanding' features, including pupil behaviour and a proportion of outstanding lessons. Teachers were said to be highly effective in using assessment information to monitor pupils' progress, which led to pupils making increasingly good progress as they move through school and achieving well in Year 6 national tests. In addition to this, Ofsted notes 'the school's ethos and atmosphere help to provide students with a well-developed sense of right and wrong and help them to be reflective and thoughtful in their actions'.

Data collected from Jane's year group at the end of Year 6 suggests CP1 Junior School was very strict and authoritarian. There was a very hierarchical management structure with strong boundaries between the headteacher, the teachers and the students. The students refer to a lack of attention to their personal and emotional needs. The regulative discourse and practice of the school was directed towards functional aspects of behaviour. This was witnessed in a series of highly directional posters displayed around the school. Most notably a number featured a picture of a dog with suggestions that students should obey commands (such as to sit down) in much the same way as a well-trained dog would do on command.

Before Jane moved to secondary school, she was interviewed about her expectations and concerns. Her concerns were coded as primarily academic. The differences between primary and secondary settings that she rated as most important were, in descending order, 'different rules', 'more advanced lessons' and 'more sets of homework'. The three most important things that would change as she entered her new school were 'the timetable', 'the lessons' and 'the teachers'. She expressed concerns about changes in instruction and regulation:

> Because when you get used to a teacher at this school they're going to be more different and you've got new rules that you've got to get used to and they're gonna set new things out and there's gonna be lots of different teachers.

When speaking about the rules of the secondary school she said she expected them to be 'stricter and more precise'. It was also clear that she considered the academic aspects of secondary school to be more important than spatial aspects. Though she referenced navigating the new school – 'It's going to be a hard way to get around the school because you're really not going to know where you're going to go' – she later compared the two aspects – 'Some people struggle with open spaces but I think that different teachers is more important than having open spaces because if there's an open space it just gives you more room but a different teacher is more of a serious matter.'

In September 2013, Jane started at C1 Secondary School, which is smaller than the average-sized secondary school with 757 students enrolled at the time of the study. The large majority of students are of white British heritage and speak English as their first language; only 2.9 per cent do not have English as their first language. The proportion of students who were supported by School Action Plus or have a statement of special educational needs was over twice the national average of 7.7 per cent at 16.8 per cent. The proportion of students supported by pupil premium was also above average at 37.2 per cent.

When Jane started secondary school, it held a rating of 'good' from Ofsted, with outstanding pastoral care and examples of outstanding teaching. The most recent Ofsted report suggested that the school has a strong emphasis on a caring, inclusive ethos. Students make good progress from starting points that are often very low and attain grades at the end of Year 11 that are average overall. Students are proud of their school and feel very safe and well cared for. The Ofsted report noted that there are very good relationships between staff and students and students behave very well both in and out of lessons.

After Jane's first week of school, she was asked to write an essay entitled 'My first week at school'. In her essay, Jane talks about her fears with regards to social relations: 'Would the older pupils hate me? Would they push me around? . . . I looked around me and saw so many pupils. There were girls and boys, short and tall ones and many more. I felt so small.' However, she later describes how helpful the older pupils were when she couldn't find her classroom. She primarily focuses on the social aspects of her transition to secondary school but it is clear that Jane was looking forward to academic changes as well: 'I had so many worries going through my head, but tried hard to push them away with good thoughts, new friends, new teachers and more to learn as well.'

At the end of the essay, she refers to being happy in her lessons and feeling settled after only one week. During Jane's first term at secondary school she completed a second NGT. In this discussion, she described the strong control and strict discipline of her primary school: 'They sort of belittled you and treated you like you was four.' She goes on to explain that she prefers her secondary school: 'Some of the teachers didn't like you to have any fun. They just made you get your head down and do some work. But here they let you have a bit of fun.' She also compared the academic aspects of her primary and secondary: 'When I was at my primary school my maths went down and I'm not sure why. I think it was the teacher where she shouted at me a lot, she helped but she'd spend more time with other students. But then when I came here my maths is starting to come up a bit.'

Her favourite spaces in her new school were her form room, the drama studio and the library. She thinks learning in open spaces is: 'Ok but sometimes when there's older students in the other classroom it does get quite annoying when you're trying to work . . . Cause if they're in like 10/11 they're shouting and all that so it gets really annoying after a while.' She does not like the glass panel walls 'because of the feeling of someone watching you through that, it feels a bit uncomfortable' and she worries about what others will think when they see her in a classroom. If she could change anything about the school, she would change the positioning of the structural columns because she sits behind one in maths, meaning her view of the board is obstructed.

When Jane was nearing the end of her first year at secondary school, she completed her third NGT. The areas she liked learning in were primarily open-plan spaces and the areas she said she disliked learning in were closed classrooms. She mentioned feeling more at ease in open spaces and having the freedom to move. She said it did not take her long to get used to the open-plan spaces, and that her class do not always share the open-plan spaces with other classes. When they do share the open-plan spaces with other classes, she says that the divider stops her from being distracted and that the noise is not too bad as people use the space conscientiously. This provides an interesting contrast as she mentioned in her NGT2 that the open-plan spaces can be loud when shared with another class. Nonetheless, Jane frequently mentions her preference for open-plan spaces in NGT3, but she acknowledges that there is a benefit of having a combination of open-plan and closed classrooms because some lessons are more suited to closed classrooms than open-plan, for example, assessments.

The only closed classroom she marked as an area she likes to learn in was her form room. This is consistent with what she said in NGT2. This probably results from the nature of activities in the room rather than the spatial features. In the form room, the activities and discussions are oriented around care and guidance rather than information and assessment. Furthermore, it is likely that Jane visits her form room multiple times a day and feels some ownership over it. This appropriation of space could easily result in a preference towards this area of the school above others. Interestingly, when asked which area of her primary school she preferred, Jane retrospectively said she preferred the outside rather than the classroom that she spent all day in.

Jane feels most observed in the heart space because teachers are usually on the balcony, but she does not mind this because teachers can see if anything happens. Moreover, Jane marked the heart space as one of the safest places in the school and one of her favourite places to be with friends. She also marked other open-plan social areas, such as the library, as the safest spaces in the school and her favourite places to be with friends, and marked the closed classrooms as the least safe and 'cramped' areas of the school as worst to be with friends. The library, the heart space and the canteen were among her favourite places

158 Changing schools

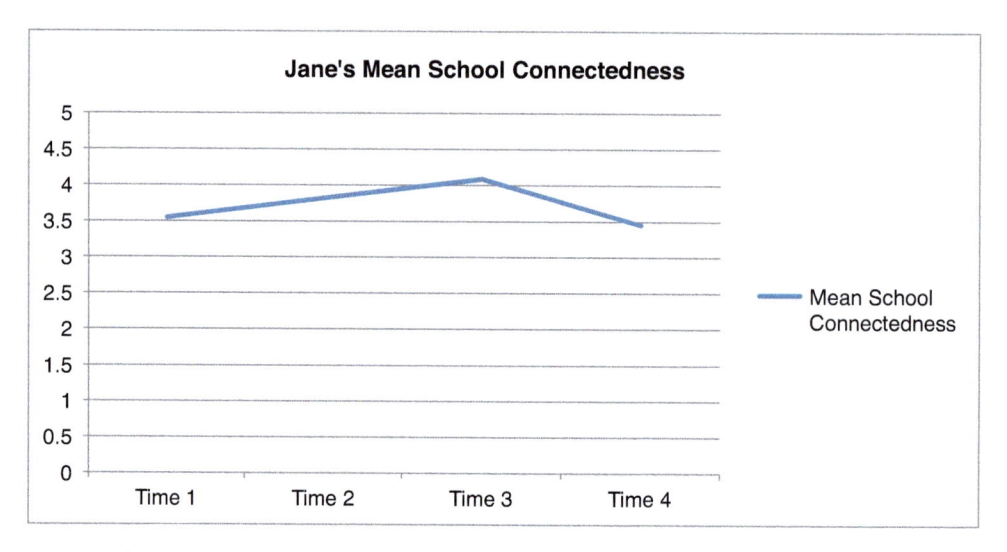

FIGURE 7.6 Jane's mean school connectedness score over time

to be with friends because they are open and she expects to find her friends there. She also nominated several outside spaces and the balcony above the heart space.

In terms of the quantitative data, Jane's transition to secondary school has been positive. Her mean school connectedness rose steadily from primary school to secondary, and only dropped slightly at Time 4 once she had been in school over a year (see Figure 7.6). This does not follow the usual pattern seen in school connectedness data where the child's scores increase when they start secondary school, but drop significantly when the 'wow factor' of a new school wears off.

John: modality B to modality A (student's name has been changed)

We now move to a compilation of data concerning the move that a student named John made from a B modality primary school to an A modality secondary school.

His primary school, DP1 (Locality D Primary School 1), is larger than the average primary school with 300 students on roll, and an above-average proportion of pupils who were eligible for pupil premium. Most pupils are of white British heritage and the proportions of pupils from minority ethnic groups and those who speak English as an additional language are below average. The proportion of students with disabilities and those who have special educational needs who were at the time the data were collected supported through school action, school action plus or have a statement of educational needs was well above average. In January 2013, Ofsted rated the school as 'good'. The report stated that 'pupils make good progress because they are taught well'.

By the end of Year 6, 'they reach broadly average standard in reading and mathematics. The Ofsted report continued to make the following observation: 'Nonetheless, pupils do not make enough progress in writing and not all pupils make consistently rapid progress throughout the school.' Teachers have high expectations of pupils and know their needs very well. Extremely positive relationships contribute to a 'stimulating atmosphere for learning'. Pupils enjoy learning because it is fun and most is challenging. Special attention is given to enhancing classrooms by providing stimulating displays, low-level lighting, background music, and interactive resources and activities designed to take account of pupils' wide range of learning styles. In lessons, pupils talk happily about their work, listening sensibly to each other's opinions, and show respect for adults. A strong feature of the school is the family spirit and the exceptional pastoral care provided for pupils. Pupils indicate that they are 'known and treated as individuals'. The headteacher's drive, energy and vision for improvement is shared by all staff. Leaders work collaboratively with staff and pupils to secure the best experience for everyone who attends the school. Equal opportunities are embedded in every aspect of school life. Pupils achieve well and all staff are supported to perform at their best.

In a formal interview the headteacher gave a strong account of a form of practice in which the regulative aspects predominated and in which boundaries between spaces were weakened, for example:

> As a school we believe providing an enriching learning environment is an essential facet of an effective school . . . we believe good displays in classrooms and around the school have a positive value because: it creates self-esteem, creates a positive and stimulating environment, celebrates work, a reference and teaching point, reflects and reinforces learning, gives a positive message and shares the ethos of the school to visitors, encourages appropriate handling of artefacts, raises standards, promotes themes, raises the profile of individual subjects . . . Year 6 is mainly taught in class but we also use school grounds for some lessons or the ICT suite for computer work . . . classroom seating is flexible according to need. Mainly groups of six children at tables but some lessons can be on floor or children find their own space.

While in the final weeks of his primary school (June 2013) John participated in a nominal group exercise. John's three most important expected differences on the voting slip were 'food' (three stars), 'harder work' (two stars) and 'PE' (one star). On the list of expected changes, 'more stricter', 'harder work', 'move between the buildings get to different lessons', 'detention' and 'food' were written down, when ordered 'food' (three stars), 'harder work' (two stars) and 'move around' (one star) were considered most important. He was looking forward to having cooking lessons in his secondary school and the new food. He was also looking forward to PE because it is his new secondary school's speciality and 'it's a fun way for staying healthy'.

Overall the nominal group of which he was a member seemed to have none of the concerns held by pupils at other Modality A primary schools. There was no mention of getting lost, no mention of bullying and hardly any negative expectations. One girl said she thought they would get to know the different teachers but if one of them did not like you, then that could be bad. There was mention of harder work but nobody seemed to mind that. This was an exceptionally well-behaved group, all the children said their names, nobody spoke over each other.

The secondary school

The school is a smaller than average-sized secondary school with 686 students on roll. The Academy is an entrepreneurial Academy with a specialism in sport. A large majority of students are white British who speak English as their first language. The proportion of pupils known to be eligible for the pupil premium is well above that found nationally. The proportions of students who were, at the time of the study, supported through school action, at school action plus or with a statement of special educational needs are well above average.

In December 2012, Ofsted deemed that the school 'required improvement'. The quality of teaching, behaviour and safety of the pupils, and leadership and management were all rated 'good' but the achievement of pupils required improvement:

> Although the achievement is improving, it is taking time for good teaching to bring students up to expected levels. Students have a lot of catching up to do because many have weak skills in reading,

writing and mathematics which limit their progress. A few subjects are lagging behind in providing consistently good teaching. The Principal and senior leaders work in close partnership with the governors and the Academy sponsors to provide excellent leadership.

In his essay, John mentions academic concerns – 'I thought everything would be really hard and tiring and boring but surprisingly it isn't, especially English and French. It's really fun and awesome.' He also describes how he enjoys music lessons because he can use lots of instruments that were not available to his at his primary school. He was not a fan of his food lessons because he has not been able to cook anything and he was really looking forward to it. He refers to spatial aspects of the school – 'I really thought I was going to get lost every lesson but so far I have only got lost once in the whole week' and 'I am gobsmacked that I work in such a big school. I am not really used seeing loads of children and adults and moving class to class is kind of hard but I think I have got the hang of it.'

In October 2013 John participated in a second nominal group during his first weeks at secondary school. John hardly contributed to the discussion. He listed his most important places as 'music', 'sports hall' and 'lunch hall' in descending order. He complained that the finger scanner does not always work so sometimes people cannot pay for their lunch. John mentions a blockade saying 'path blocked', which means they have to go a certain way into school. The school controls student movement very strongly.

The whole group discussed the danger of people entering the school through the gates. There is no agreement on whether they prefer the gates at BACA where you have to have a card or be let in/out by the teachers, or the gates at DP1 where you have to press a button.

This was not a very well-behaved NGT group. The children spoke over each other and whispered to their friends throughout. When told that they can have a drink, they seem to have a fight at the drinks table. This form of behaviour is in stark contrast to the primary school NGT. These were the same students who were so well behaved in the primary school NGT.

John's mean school connectedness is interesting as it follows the opposite trajectory to the overall average school transition. Usually, the school connectedness score increases at Time 2 and decreases at Time 3. John's mean school connectedness decreases at Time 2 (from 3.45 to 3.18) then increases at Time 3 (to 4) (see Figure 7.7). Moreover, the particular items that decreases were 'proud to be part of the school'

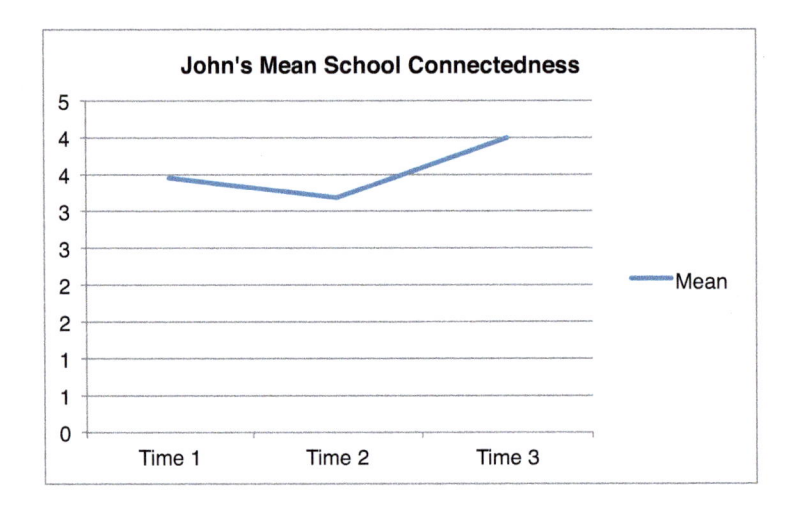

FIGURE 7.7 John's mean school connectedness score over time

(from four to three), 'have time on my own' (from three to two), 'easy to find way around' (from five to three) and – most dramatically – 'able to be myself' (from five to two). Nonetheless, these items increased again at Time 3 so perhaps the contrast of the two schools resulted in more extreme answers.

Discussion and conclusion

The data reveal that initial reactions to secondary school are significantly affected by the primary school experience, and that there is a general tendency for feelings of school connectedness to decline by the end of the first secondary year in all types of school. A good deal is revealed about students' concerns around this time. These are largely around personal well-being and social relations, with fears about safety and isolation paramount. The data give a strong insight into the role design plays in relation to these concerns, with particular importance given to the balance of private and public spaces, and the attendant security issues (e.g. concerning design of toilet areas), issues of security and surveillance (with a need for balance between making children feel safe because visible and insecure because over-seen), and logistical issues including ease of navigation and access to facilities and resources. Although overall the BSF-type designs had an initial positive effect on feelings of school connectedness, this fell away during the first secondary year. However, as with all the data, the differences between schools are considerable, and are strongly driven by management practices.

Recent policy on school design promoted transformations of modalities of design and pedagogic practice. In the sample of schools that we studied, some of these transformations were maintained over time while others were repeatedly adapted, in sometimes vain attempts to recreate prior modalities of design and practice.

We used multiple sources of interview and observation data to populate a model of description based on Bernstein's work. There was a very high degree of consistency over these multiple sources. We were also able to identify schools where the design vision was aligned with current practice and where the modality of current practice was in direct conflict with the modality of the school design. On the basis of these data, we were able to identify four types of trajectories between two modalities of primary schools and the modalities of secondary schools.

Our internally consistent adaptation of the school connectedness scale showed the decrease over time that is characteristic of this measure (Niehaus et al., 2012). However, when the contrast between the new-build schools and established comparator schools was examined, the decrease in the new-build schools were significantly less in time point 2 at the beginning of secondary school (see Figure 7.4), although there was no significant difference by the end of the first year at secondary school. When we discussed this finding in schools, teachers spoke of a 'wow' effect that they felt 'worn' off over a short period time. Some schools went to considerable effort to maintain this 'wow' effect as long as possible through such measures as constant repainting of scuffed surfaces. A national comparative study of attainment in new-build schools also showed a similar effect (Williams et al., 2015). There are strong claims for an improved behaviour effect in new-build schools which we found support for in interviews with headteachers (PWC, 2003):

> The behaviour has improved dramatically, the standards of teaching and learning, the standards of student behaviour, the standards of progress and attainment in our school over the last five years has shot up astronomically.
>
> *(Headteacher A1 interview)*

When the connectedness data were examined at the school level, we saw that changes in connectedness scores differed markedly over time between schools. The highly significant differences between trajectories (see Figure 7.4) that had an A-type secondary school and those that had a B-type secondary school were supported by extensive interview and observation data which will be reported in Chapter 8. This was particularly marked in cases where students from B-type primary schools moved to A-type secondary schools.

However, it was only when the data were examined over time between school modalities that significant group effects were observed. Additionally, comparisons between schools where practice was aligned with design and where practice was not aligned with design show highly significant differences in the secondary school measures where there were no significant differences in the primary data.

In answer to our overriding question as to whether design matters, it seems that the relation between design and practice is what matters, as well as continuity in the experience of design. The internal coherence of design and practice also seems to be important. Perhaps of more interest is the evidence we have obtained on the implications of moving from one modality of design and practice to another. It would seem that school design is one of the factors that act as explicit and tacit relays of the structure of pedagogic practice and that when students move between different structures of pedagogic practice over time, they are faced with specific challenges in adjusting to what may be thought of as new semiotic orders in which specific forms of competence are privileged. When we followed students from the end of Year 6 in their primary schools through their transition into Year 7 of their secondary schools, we were examining the implications of moving from one pedagogic environment to another and one physical/design environment to another. Our concern was the effects of continuities and discontinuities in these trajectories. We have shown that this does matter. Our focus was on the effects of change in physical and pedagogic factors at the institutional level as experienced at the personal level by students.

Hundeide (1985) has shown, in a study of the tacit background of children's judgements, how participants in an activity, in part, create the setting. These 'taken for granted background expectancies' reflect in part the sociocultural experience that the individual brings to the situation:

> One needs a framework that takes into account the historical and cultural basis of individual minds: the collective institutionalised knowledge and routines, categorisation of reality with its typifications, world view, normative expectations as to how people, situations, and the world are and should be, and so forth. All this is tacit knowledge that has its origin beyond the individual, and it is this sociocultural basis that forms the interpretive background of our individual minds.
>
> *(Hundeide, 1985, p. 311)*

Bernstein's (1981) chapter outlined a model for understanding the construction of pedagogic discourse. In this context pedagogic discourse is a source of psychological tools or cultural artefacts:

> Once attention is given to the regulation of the structure of pedagogic discourse, the social relations of its production and the various modes of its recontextualising as a practice, then perhaps we may be a little nearer to understanding the Vygotskian tool as a social and historical construction'
>
> *(Bernstein, 1993, p. xviii)*

He also argues that much of the work that has followed in the wake of Vygotsky 'does not include in its description how the discourse itself is constituted and recontextualised':

> The socio-historical level of the theory is, in fact, the history of the biases of the culture with respect to its production, reproduction, modes of acquisition and their social relations.
>
> *(Bernstein, 1993, p. xviii)*

As Ratner (1997) notes, Vygotsky did not consider the ways in which concrete social systems bear on psychological functions. He discussed the general importance of language and schooling for psychological functioning, however, he failed to examine the real social systems in which these activities occur and reflect. Vygotsky never indicated the social basis for this new use of words. The social analysis is thus reduced to a semiotic analysis which overlooks the real world of social praxis (Ratner, 1997). Vygotsky's understanding of mediation by psychological tools is, as it were, situated by the Bernsteinian understanding of the regulation, structuring and recontextualisation of the tool. In this way a psychological understanding of the social formation of mind is extended through a sociological understanding of the origins of mediational means.

In a recent chapter, Singh (2017) compares Foucault and Bernstein's ideas on governance. She suggests that new policies on school design are in effect new modes of pedagogic governance, and these modes of pedagogic governance are recontextualised in specific practices. Our work has shown how new policies of school design have been recontextualised (Daniels et al., 2017) as understood by Bernstein (see Singh et al., 2013). Attempts to effect the transformation of education as outlined in BSF depend on the willingness of schools to align with an imagined pedagogic future. It also depends on the availability of the psychological tools (Vygotsky, 1987) which enable the capacity to bring transformation into effect.

> The allure of continuous makeover by pedagogic means is supposedly available to all actors. But this is the fantasy of pedagogic makeovers. Implicit within the new pedagogic translations is a model of the ideal learner who has the capacity (not ability) to '*meaningfully* rather than relevantly or instrumentally project' themselves into a pedagogised future.
>
> (Bernstein, 2001, p. 366, original emphasis, in Singh, 2017, p. 159)

This project advances the development of the post-occupancy evaluation of schools through the incorporation of perspectives drawn from Vygotsky's theory of sociogenesis and Bernstein's later work on cultural transmission. The chapter has also shown how Bernstein's approach to the codification of modalities of pedagogic practice can be extended to incorporate a broader notion of the configuration of space in the design of a building and allows for the examination of the consequences of change over time.

We suggest that innovations in school design must be understood as relays of underlying arguments that may come into conflict with other pedagogic perspectives in the social world of schooling. The interplay between design and practice can ease or exacerbate the challenges of moving between schools.

References

Abreu, G. and Elbers, E. (2005) The social mediation of learning in multiethnic schools (introduction), *European Journal of Psychology of Education*, 20(1): 3–11.

Benito, A.E. (2003) The school in the city: School architecture as discourse and as text, *Pedagogica Historica*, 39(1–2): 53–64.

Bernstein, B. (1977) *Class, Codes and Control, Vol. 3*. London: Routledge and Kegan Paul.

Bernstein, B. (1985) On pedagogic discourse, in J. Richards (ed.), *Handbook of Theory and Research for the Sociology of Education* (pp. 205–40). Greenwood Press: New York.

Bernstein, B. (1990) *Class, Codes and Control, Vol. 4: The Structuring of Pedagogic Discourse*. London: Routledge.

Bernstein, B. (1993) Foreword, in H. Daniels (ed.), *Charting the Agenda: Educational Activity After Vygotsky* (pp. xiii–xxiii). London: Routledge.

Bernstein, B. (2000) *Pedagogy, Symbolic Control and Identity: Theory, Research, Critique*, rev. edn. Lanham, MD: Rowman and Littlefield.

Building Bulletin (BB) 95 (2002) *Schools for the Future: Designs for Learning Communities*. Department for Education and Skills.

Burke, C. (2010) About looking: Vision, transformation and the education of the eye in discourses of school renewal part and present, *British Educational Research Journal*, 36(1): 65–82.

Carvalho, J. and Dong, A. (2006). Legitimating design: A sociology of knowledge account of the field, *Design Studies*, 30: 483–502.

Cole, M. (1998) *Cultural Psychology: A Once and Future Discipline.* Cambridge, MA: Harvard University Press.

Daniels, H. (1989) Visual displays as tacit relays of the structure of pedagogic practice, *British Journal of Sociology of Education*, 10(2): 123–40.

Daniels, H. (1995) Pedagogic practices, tacit knowledge and discursive discrimination: Bernstein and post-Vygotskian research, *British Journal of Sociology of Education*, 16(4): 517–32.

Daniels, H. (2010) The mutual shaping of human action and institutional settings: A study of the transformation of children's services and professional work, *The British Journal of Sociology of Education*, 31(4): 377–93.

Daniels, H., Tse, H.M., Stables, A. and Cox, S. (2017) Design as a social practice: The design of new-build schools, *Oxford Review of Education*, 43(6): 767–87.

DfES (2003) *Classrooms of the Future: Innovative Designs for Schools.* London: DfES.

Douglas, J. W. (1964) *The home and the school: a study of ability and attainment in the primary school.* MacGibbon and Kee: London.

Fitz, J. (2007) Review essay: Knowledge, power and educational reform, applying the sociology of Basil Bernstein, *British Journal of Sociology of Education*, 28(2), 273–279.

Flygt, E. (2009) Investigating architectural quality theories for school evaluation: A critical review of evaluation instruments in Sweden, *Educational Management Administration and Leadership*, 37(5): 645–66.

Goodenow, C. (1993) The psychological sense of school membership among adolescents: Scale development and educational correlates, *Psychology in the Schools*, 30: 79–90.

Hasan, R. (2002) Semiotic mediation, language and society: Three exotropic theories – Vygotsky, Halliday and Bernstein. Presentation to the Second International Basil Bernstein Symposium: Knowledge, pedagogy and society, Cape Town, July.

Hoadley, U. (2006) Analysing pedagogy: The problem of framing. www.education.uct.ac.za/sites/default/files/image_tool/images/104/hoadley2006.pdf

Hundeide, K. (1985) The tacit background of children's judgments, in J. Wertsch (ed.), *Culture, Communication and Cognition* (pp. 306–22). Cambridge: Cambridge University Press.

Ivic, I. (1989) Profiles of educators – Lev S. Vygotsky (1896–1934), *Prospects*, XIX(3): 245–58.

Luria, A.R. (1928) The problem of the cultural development of the child, *Journal of Genetic Psychology*, 35: 493–506.

Melhuish, E.C., Phan, M.B., Sylva, K., Sammons, P., Siraj-Blatchford, I. and Taggart, B. (2008) Effects of the home learning environment and preschool center experience upon literacy and numeracy development in early primary school, *Journal of Social Issues*, 64(1): 95–114.

Moos, R. H. (1979) *Evaluating Educational Environments: Procedures, Measures, Findings and Policy Implications.* San Francisco, CA: Jossey-Bass.

Niehaus, K. Moritz Rudasill, K. and Rakes, C. (2012) A longitudinal study of school connectedness and academic outcomes across sixth grade. *Journal of School Psychology*, 50: 443–60.

PricewaterhouseCoopers (2003) *Building Performance. An empirical assessment of the relationship between schools capital investment and pupil' performance.* DfES Research Report Series, RR242.

Ratner, C. (1997) *Cultural Psychology and Qualitative Methodology. Theoretical and Empirical Considerations.* London: Plenum Press.

Singh, P. (2017) Pedagogic governance: Theorising with/after Bernstein, *British Journal of Sociology of Education*, 38(2): 144–63.

Singh, P., Thomas, S. and Harris, J. (2013) Recontextualising policy discourses: A Bernsteinian perspective on policy interpretation, translation, enactment, *Journal of Education Policy*, 28(4): 465–80.

Tudge, J.R.H. and Winterhoff, P.A. (1993) Vygotsky, Piaget, and Bandura: Perspectives on the relations between the social world and cognitive development, *Human Development*, 36(2): 61–81.

Vygotsky, L.S. (1987) *The Collected Works of L.S. Vygotsky. Vol. 1: Problems of General Psychology, Including the Volume Thinking and Speech*, ed. R.W. Rieber and A.S. Carton, trans. N. Minick. New York: Plenum Press.

Wertsch, J.V. (1998) *Mind as Action.* Oxford: Oxford University Press.

Williams, J.J., Hong, S.M., Mumovic, D. and Taylor, I. (2015) Using a unified school database to understand the effect of new school buildings on school performance in England, *Intelligent Buildings International*, 7(2–3): 83–100.

Credit: Tim Crocker

8
WHAT MATTERS ABOUT DESIGN?

Introduction

In the preceding chapters, we have described and discussed the activities and findings of the *Design Matters?* project. We have discussed the ways in which the discourses and practices of school design produce educational spaces which influence and are influenced by the discourses and practices of teaching and learning when the building is occupied. We have also discussed the development of a methodology for systematically analysing the relationship of school space to the experiences of students, teachers and parents. This expands notions of post-occupancy evaluation (POE) research by exploring how the motives of an educational vision which informed an initial school design, those of the final build and those of the people who occupy that building interact in a way which influences experiences of the end-users. In this way we sought to understand more about the extent to which a building regulates or governs the behaviour of those who occupy it. Through our approach to multi-professional post-occupancy evaluation we have come to the view that a building may be understood as a tool which may be used to facilitate change rather than as an instrument of change in its own right.

In this final chapter, rather than presenting an account of the more systematic aspects of the project, we will draw on some of the complex issues that have come to our attention as 'noticings' alongside the more formal aspects of data collection and analysis. We have found that these matters that we have noticed have formed rich points of departure for our thinking about the future of research in this complex and challenging area. We present each of these 'noticings' as discussion points.

Discussion point 1: the technological 'future'

By way of a point of departure, we will first consider the roots of some of problematic issues of the initiative which we studied. In part, *Design Matters?* was prompted by the unravelling of the *Building Schools for the Future* initiative. There was widespread, often highly politically motivated, concern about the returns on such a high level of investment in refreshing the school estate.

Our attention was caught by two particularly problematic dimensions. The first of these was with respect to predictability. If such a large amount of money was to be spent on building schools for a future, then how predictable was that future? For example, considerable amounts of time and energy were expended in thinking about how new technologies could exert a transformative effect on practices of schooling and how schools should be designed to facilitate the development of technology-mediated practices of teaching and learning. However, at the time of envisioning, designing and constructing these new schools the future shifted.

The original 'future' was one of a deployment of technology in schools that was desktop. The 'breakout' spaces in our sample schools were sometimes designed as workstations with fixed computers, or leads for computers, for individual or joint work. These elements have now been removed. The technology has changed and the emphasis is now on mobile wireless access to systems of information retrieval, communication and analysis. (There are other reasons for the general lack of utilisation of breakout spaces which we discuss below.)

In at least one of the sample schools, pedagogic changes since our period of formal data collection have moved the school even further from the original technological vision, with students banned from any use of mobile technologies in school, on the grounds that the school and examination system remains dominated by paper-and-pencil technology, so the school needs to act as a counterforce to the external world of digital media. This might seem an extreme abreaction, but this policy is enacted on behalf of an academy chain brought in to improve the school's performance in relation to current criteria. The world does not always move as predicted! (Consider, for example, the likely fate of a 'Curriculum for the Twentieth Century' produced in 1910.)

Discussion point 2: demographic changes

There have also been significant demographic changes with a rise in the demand for school places and changes in ethnic mix. This has brought with it concerns about whether the use of space as imagined in BSF was most appropriate in an era of constrained financial resources and limited capacity for expansion. There are major concerns about whether the vision of pedagogic and spatial flexibility as well as adaptability over time envisioned in BSF were fit for a more austere future.

We noticed three dimensions of this problem. The first relates to school population and classroom size. Perhaps more out of hope than expectation (though we cannot be sure), the schools we investigated were often large buildings with relatively small classrooms. There was limited capacity to house classes of 30 or more. This limitation brings with it, of course, significant challenges to budgets if, say, seven teachers need to be employed to teach a year group of 150 at any one time, rather than five.

There have been marked changes, too, in the ethnic mix of some of the schools. This was often most apparent in formerly 'white working-class' areas, where the school population is now much more diverse. This has brought two specific issues for schools. The first relates to the demand for English language support. Again, the original BSF vision did not cater strongly for this need to be met, and headteachers shared with us their frustration at lacking appropriate spaces for this work to be carried out. A second issue relates to religion: this was not something that featured strongly in the BSF vision, but some schools struggle to find appropriate spaces for prayer rooms, notwithstanding any other concerns in this area.

It is interesting to consider the prayer room issue in relation to a broader issue around BSF schools: that of visibility and security *vs* personal space. As noted elsewhere, the new designs made it very difficult for students – even staff – to find somewhere quiet, private and relatively hidden. It may be that the advantages of this outweigh the disadvantages. On the other hand, there was some evidence of a desperate seeking of peace and privacy, as in the cases of some pupils locking themselves into toilet cubicles to eat lunch.

Discussion point 3: policy aspirations and joined-up thinking

A major issue that seemed particularly problematic was with respect to the effectiveness of an intervention that operates in one dimension. In our case, of course, this is design. As we argued in Chapter 2, the BSF initiative was launched in a flurry of claims for the transformation of schooling. However, an intervention that prepares for a future cannot operate in one dimension. The envisioned practices of BSF carry with them implications for curriculum, assessment, teacher training and the inspection system that were not either maintained or realised, and may not have even been attempted in some cases. The arguments in favour of design for personalised pedagogic practice do not sit comfortably with current priorities and may not with those that emerge in the coming years.

A key indicator of this in our own research related to the use of breakout spaces. The BSF schools were designed to promote more flexible and personalised teaching and learning. However, our extensive classroom observations revealed very little pedagogical use of these spaces, with only a significant amount of such use in School B1where, as noted previously, there was an unusual level of continuity between from the original vision to the current school management. Indeed, in the vast majority of cases, the lessons were teacher led, entirely classroom based, and not obviously different from what we would expect to see in any school, however old – albeit often carried out in very well-appointed and well-lit classrooms.

For this to be otherwise would have required a number of changes beyond BSF, that is, beyond merely 'building' schools for the future. Teacher training and the inspection system would have had to have been geared towards encouragement of allowing students to leave the classroom to work relatively unsupervised, for example. The curriculum and assessment systems would have to have moved to validate and encourage such work.

That these changes did not come about can be attributed to a range of factors. One of them being the emergence of a much stronger child protection agenda.

Discussion point 4: child protection and the open community school

The BSF schools were built to be strongly outward-facing, allowing the external community relatively free access and generally breaking down the barriers between school and broader community. Indeed, community regeneration was a strong motivator of the policy.

During our data collection period (though there has been some rowing back in some schools since) there was a fierce reaction to this, on the grounds of child protection. Far from allowing free access, these schools were some of the most security conscious we have visited, with numerous security gates having been added to the original design. Even within the schools, there were examples of previously free movement being restricted. This can be seen as a conflict of two equally desirable aims, where one takes priority over the other at a particular point in history, and where some trade-off may be inevitable. This notwithstanding, the strengthening of the child protection agenda served to undermine the original design intention of openness to the community.

Discussion point 5: flexibility, adaptability and the principles of design

The extent to which the outcomes of the BSF initiative represented a sufficient level of flexibility and adaptability required for the lifecycle of a building which will be in operation against the backdrop of changes in pedagogic fashion is a moot point. This issue may be thought of in terms of an understanding of 'sustainability', which can be thought of as much more than a matter of environmental costs of building or running a school. There is also the question of the extent to which the design is sufficiently flexible and adaptable to respond to changing demands either as a result of changes in leadership or policy. In Chapter 6 we illustrated the radical shifts in demand that can take place with change of headteacher and as we saw, these changes can take place surprisingly regularly.

In Daniels et al. (2018) we reported on some of the *Design Matters?* findings which reveal some of the ways in which school buildings play a role in mediating the pedagogic process. We also showed how contradictions embedded in the design of a building and the building process itself help to shape the possibilities for pedagogic practice which in turn may also seek to re-shape the building itself. It is the tensions that are set up between these strands of development which have given us insight into the way in which mediational processes progress after occupation.

Our data suggest that some buildings may be so riven with contradictions that adaptations to particular preferences may prove ineffective and the building becomes seen as dysfunctional. This may either be because of features internal to the design or because of relations between practices of construction and funding. We have evidence of adaptations which were successful in re-shaping these school buildings in a way that rendered them more fit for the purposes of the occupiers.

The view of BSF as an act of social engineering whereby new designs would invoke changes in practice did not allow for the agency of occupiers:

> These buildings have been designed in a way that makes it almost impossible for them to go back to square one. That was our guiding principle.
>
> *(Local County Council Transformation Team Member)*

We have discovered that schools are rather sites of practice than determiners of practice. That is to say, buildings can offer new occupiers certain invitations and certain constraints, but each new occupier comes with a strong intention to undertake certain activities in certain ways. These intentions are likely to be modified, but are unlikely to be overturned, by the constraints of the building.

Our data suggest that the relation between design and practice is crucial to the production of a building which can be and is used effectively. The suitability of the building for a school's pedagogic practices as they change through time will be determined by the building's potential to adapt to the school's changing spatial needs and the school's understanding of the building's design principles. There are three elements to this relationship. First, it is more likely that a successful occupation and use of a building results when the practices that the occupying staff wish to follow or mirror are the principles of practice that are embedded in the vision and design. Second, this is most evident when the eventual practitioners (usually the headteacher who takes over the school building on completion) have been involved in an inclusive consultation process throughout the vision, design and construction process. (We should stress, however, the very high turnover rate of occupiers among the schools we studied, such that the influence of the first headteacher may be extremely diminished.) Third, it is quite clear that the principles of the design brief may be regarded differently by different individuals and professional groups. This may seek to compound problems with the relationship between design and practice.

These conclusions lead us to form a general argument that one design may be perceived and used in very different ways in different practices of schooling. We also argue that good design requires good multi-professional holistic post-occupancy evaluation which has a remit that goes far beyond the physical functioning of the building. An understanding of social relations that are enacted within a design as it is taken up by different forms of practice is crucial to the development of better sites for schooling. These social relations comprise both the formal aspects of the teaching-and-learning relationship and the informal aspects of student – student and student – staff interaction, safety and well-being.

In short, we suggest that different patterns of collaboration and mutual shaping of design and practice are vital elements in the processes through which a school is designed, built and occupied. This requires clarity and continuity in the operational definition of the client. In an interview a senior architect suggested that the key role of the architect was as 'orchestrator' and a contractor suggested that 'integrity must lie at the heart of a build'. In their different ways they appear to recognise the need for the formation of common objects of the work. However, it is not clear that our interviewees all fully acknowledged the extent to which usage can undermine design intention.

We can take a non-educational example to show how building design can accommodate quite contrary practices, though in this case we assume the outcome is intentional. A new airport terminal building is designed as a large, open, flexible space but it incorporates strict controls and restrictions in use, such as the temporary barriers used to channel and monitor queues at check-in and security. A building without internal borders certainly does not necessarily imply internal freedom of movement. We saw this clearly in some schools, where pupil movement was tightly controlled within relatively unstructured internal spaces.

Importantly we have shown how one design can be used in different ways. Rather than design determining behaviour, it takes up a dynamic and fluctuating relation with the practices of occupiers resulting in a wide variety of outcomes. Different approaches to school leadership and management give rise to distinctive school cultures which in turn make differences in the use and adaption of a school building. We take the findings as a strong argument for the development of a social and cultural dimension to post-occupancy evaluation which examines human practices in buildings over time and through different management cultures. It is as if there is a process of resignification at each point of cultural change in successive management regimes. As the headteacher of a successful new-build free school noted:

> The design is a provocation to learn differently but it's what you do inside it that matters.

Discussion point 6: the broadening of post-occupancy evaluation

Taken together, these findings point to the need for post-occupancy evaluation that includes human action and perception over time and the inter-connection between design and practice and how this may change over different occupations (school leaders). In forms of work that are as complex and nuanced as design, construction and occupation, there is a need for a more sophisticated model of description that can generate a range of possibilities for collaborative action across agencies, places and time.

The findings also point to the need to redefine 'sustainability' in terms of adaptation to different forms of practice. In order to extend the functional life of new school buildings, the vision and design process must allow for adaptation as educational policies and practices change through time.

Thus the original assumption that design could exert a direct causal influence on practice was not witnessed in the observations we undertook. This was simply a matter of headteachers not sharing original design intentions and vision with the priorities of the original design, construction and occupation.

In Locality D we were alerted to different forms of rejection of the original design principles. We were made aware that many of the students felt uncomfortable with the intentions of the original vision. One student remarked:

> This school is too posh for us.

This is a difficult remark to interpret. It is an example of one the matters we *noticed* but on which we did not have sufficient data to make a general claim. In the same way that sociologists of the 1970s and 1980s (e.g. Sharp and Green, 1975) discussed working-class rejection of pedagogies that were seen by some students to indulge the predilections of the middle class, so it may be that for students from disadvantaging social positions may perceive certain forms of design and a form of class-based imposition.

The headteacher in this school spoke of the difficulties that staff, students and families had in 'coming to terms with' the new school:

> Part of the academy was about really changing attitudes and mindsets and you're either on the bus or you're not on the bus. Some are still hanging on by the bars!
>
> *(Headteacher interview)*

The headteacher also remarked that it was impossible for student football teams to visit the school as students would engage in stone throwing:

> Perceptions are changing, because of the building . . . that positivity is starting to feed out into the community, and people are wanting to come here. Whereas in the past you wouldn't get anybody from across the city coming to play football here, they wouldn't play these children at football because they'd have stones thrown at them.
>
> *(Headteacher interview)*

All these indicators of disaffection and dissatisfaction were seen to ease over time. However, teaching staff continued to resist the visibility design intention of the original vision. Many of the classrooms had floor-to-ceiling glass walls. Visibility was reduced by the judicious use of posters and pictures which were attached to glass so as to minimise visibility. Staff suggested that they were seeking the privacy that was denied by the principles of high visibility built into the design. They also informally resist the configuration of space by creating ad hoc boundaries within large open areas through the deployment of furniture as impromptu walls. Not only were these moves impromptu, they were also highly ineffective. In several schools we noticed these kinds of arrangements of furniture that barely sufficed to afford a physical boundary. They were utterly ineffective as acoustic barriers. The original design vision was that of a large area with breakout facilities in which 90+ students of the same age would work on the same curriculum topics. However, when the area was divided by furniture and classes of students of different ages were taught different subjects, then the acoustic flow from one class to another was a distressing experience for many students. Referral to Cooper's (1985) study which examined similar problems with ad hoc boundaries and discursive marking of teacher's territory serves as a reminder of the brevity of what may be termed the 'institutional memory' of design research. All these issues should be picked up by improved forms of post-occupancy evaluation.

Discussion point 7: belonging, engagement and appropriation

These issues serve as a reminder of how transformations of space can affect a personal sense of belonging and how different forms of belonging may arise as the relationship between design and practice shifts over time. This is as much true with respect to external relations with the wider community as it is for teachers and students in the school.

It is commonplace to distinguish between 'space' and 'place'. A space becomes a place when it becomes a form of human dwelling. We suggest a third point on this continuum: 'home'. A building feels like a home when it is appropriated in a certain way by its users. (In relation to this, we noticed a general tendency for primary school classrooms to be far more strongly 'appropriated' by their students than secondary school classrooms, and we also note the prohibition of such things as wall posters in some BSF schools.)

Let us take the example of a school science laboratory. This is designed to have the appropriate elements in the right arrangement: work benches, sinks, etc. Only when it is used can it be evaluated as a place rather than a space. However, the real aim is not only that it should function as science laboratory but also that it should be a place in which students feel comfortable and empowered: a place that inspires them to study science. Now, it is of course difficult to determine how far design per se can make a space feel like a home (notwithstanding the conflating of the terms by the house-building industry), but any effective evaluation of a school design long term should surely have this element in mind. In short, how well does the building work in achieving its educational aims?

In this respect, as in others, intentions are not outcomes. In the school A1, the first headteacher we interviewed talked about design for parental engagement and participation. She encouraged the architects to design the entrance to the school as a 'funnel' which would welcome visitors into the building. Her priorities were cast in terms of forming supportive partnerships with parents which would lead to open and honest discussions about student progress and needs. She spoke of the need to focus on emotional needs as well as academic progress in these discussions and of design features that would provide a comfortable setting for such discussions. The entrance to the school was designed to bring visitors and students straight into a dining area, which was referred to as a 'restaurant'. At the start of the school day this restaurant provided tea and toast free of charge. We were told that the bread was donated by a local bakery. It was in this 'breakfast' setting that parents staff and students could share refreshment and it was in this setting that conversations could take pace with members of the community who had previously been regarded as hard to reach:

> I think the message that it gives now is the openness and the welcoming, and the non-fear factor for them. And I think those things have been the most crucial things that have enabled us to work with parents. We want to be open to the community, so in other words we want to allow parents in, in the morning, into the restaurant and very much use the space to be able to do things in.
>
> *(Headteacher interview)*

This openness to parental engagement in the practice of schooling was discussed in terms of a learning through participation model (e.g. Lave and Wenger, 1991). Similar arguments were made with respect to movement through the school by visitors and students. The headteacher spoke of her intention that the school should arrange the pedagogic circumstances in such a way that would facilitate the supposed acquisition of competence by students in organising themselves in time and space with free (unregulated by bells) movement around the school:

> I think the most pleasure about the building is that people are continually moving around school. It's not one of those school that you come into and it's like, yes the bell was rung outside and the pupils were rushing, it's time now to go to your lessons in a very orderly way. We've always said we were about opening up to the community, letting parents in, making sure the pupils have that free flow to the restaurant etc., break times, lunchtimes, it's not that either you're in the restaurant or you're outside, it's all of that free flow around.
>
> *(Headteacher interview)*

The positive encouragement on the part of the school for parents to participate in evening classes in particular areas of the school without closure of the rest of the space also rested on similar assumptions. The school assumed that encouragement to act in a positive manner was more effective than 'policing' negative action. For example, signs directed visitors to the location of evening classes without the perceived need to lock or close off other areas of the school.

These initiatives were not made in the absence of a concern for control. Rather it was a matter of mode of control that was more personalised and pervasive. This was witnessed in the 'welcoming' of students into the school in the morning albeit with 14 members of staff attempting to personalise control both at the school gate and in the entrance hall. Students who appeared to be over excited, or distressed, were engaged in conversation by staff who sometimes continued conversations as students made their way into the main body of the school.

In these ways satisfactory movement through the larger, more open areas of the school and between sections of the school was seen as a form of competence that was acquired through participation in forms of practice based on high levels of encouragement for positive and acceptable behaviour.

This system was put under extreme pressure when a parent made an unwelcome and aggressive entrance into the school. The response was to instigate a shift to more explicit modes of control and the retrofitting of electronically controlled barriers between sections of the school which required electronic keys held by staff and not students for access.

The school GCSE results reported in the summer of our first year of engagement with the school were disappointing and the headteacher decided to resign before the start of the new academic year.

The change of headteacher resulted in radical changes in modes of control and management in the school. Movement into and within the school was the subject of highly explicit forms of regulation. Boundaries were introduced at the school gate, at the school entrance and between areas of the school. In effect, the design of the school was adapted to ensure a much higher degree of overt control on patterns of participation and engagement. The offer of free toast and tea was withdrawn. Parental access to the school was restricted and the expectation became that they would not venture beyond the school gate.

This case provides an interesting example of shifts in design and practice over time, shifts that were not envisaged throughout the design process. It also points to the necessity of some form of pedagogic post-occupancy evaluation. The ways in which power and control are exercised and impact on design and practice could be at the core of such an evaluation. If the notion of boundary is thought of at a very general level, then it may be understood as way of distinguishing 'what is and what is not'. In terms of the curriculum, the relationship between history and geography may be marked very explicitly in the formulation of content and the exercise of the timetable. Alternatively the power relation may be weakened, the boundary softened and a cross-disciplinary initiative may emerge. Similarly the distinction between specialised spaces may be marked by walls, doors and barriers. These may be further strengthened with locks and regulations for use stating who may or may not enter. Our work suggest that when pedagogic boundaries are not aligned with designed boundaries, then tensions emerge in the practices of the school. Furthermore, designers and policymakers should acknowledge the inevitability of future occupiers departing from the aims behind the original vision, unless schools are only to be built to stand for a very small number of years. Contradictions between pedagogy and design are to some extent inevitable. They may both direct and deflect the attention of teachers and students away from what should be shared goals. As shown above, the transformation in forms of access for parents is one such change that carries design implications.

Problems emerging over time

In the *Design Matters?* study, there was a spectrum from coevolution to contradiction in the way that use of buildings developed and were transformed over time. A full understanding of how design works cannot be gleaned form an evaluation of its physical or environmental functioning. What is required is an evaluation of design in practice and practice in the design.

There is, of course, the matter of the practice of design which was the focus of the first research question of *Design Matters?* In Tse et al. (2014) and Chapter 3 of this book we discussed the internal dynamics between different players and agencies in the processes of envisioning, designing, constructing and occupying a school. There are important issues with respect to the motives at play in these processes or more specifically the lack of alignment in the motives. If the objects which motivated the activities of the different professions and stakeholders are not aligned, then as we have shown a building which is not fit for purpose may well result.

This issue is perhaps most visible in the processes of value engineering and cost saving which take place following the award of a construction contract and prior to handover to the client. Many of the issues we have already discussed may be thought of as lying at the macro scale. For example, the non-inclusion of sound dampening acoustic materials in what became classroom settings which were far too noisy to allow for effective teaching and learning to take place. The root of the contradiction that gave rise to the non-inclusion of these materials was made manifest in interviews with a constructor and an acoustic engineer. Prior to the award of the contract the constructor was driven by a motive of securing the contract. To this end he needed to assure the client that the building would be acoustically fit for purpose. The company hired an acoustic engineer who was of the view that this could not be achieved with the mixed economy of space design and the building materials that had been selected. The outcome of an extended negotiation was that the construction specification stated that the company would work towards the required acoustic standard in the full knowledge that this could not be achieved. On the award of the contract the constructor was driven by a new motive of bringing a build in on time and on budget. The result was that the acoustic panels, which would have made some difference to the acoustics of the teaching areas, were not included in the construction for economic reasons. The result is a level of acoustic performance which teachers and students find very difficult. They attribute their feelings of dissatisfaction to the design rather than to the design as constructed. This distances from the pedagogic ideas that underpinned the design.

This raises the thorny issue of what is a form of competence (in the Hymesian sense) and performance tension. When a question such as 'does design matter?' is raised, it can bring an assumption that statements can be made about the value of a particular design approach. This matter of acoustics points to the issue of how well the design has been constructed. A poor-quality construction and a high-quality construction of the same design may be evaluated in very different ways (Price et al., 2009).

At a more 'micro' level the issue of cost saving may be seen to make and impact on practice. In School B1 the constructors purchased enough taps to fit all the basins as an 'end of line' transaction. These discontinued taps cannot now be repaired when they fail. Clearly, this issue does not make a direct impact on the practices of teaching and learning in the school. However, it does detract from the aesthetics of the school. This matter of what is often referred to as the 'wow factor' is a matter of much concern to many of the heateachers we interviewed. In one school there was a budget allocation for white paint in order to preserve the pristine appearance of the classrooms and corridors. Corridors were cleaned after every breaktime in order to remove any unslightly rubbish and every year worktops were sanded down and revarnished. As in several schools that we visited, an argument was made that if students felt that they were being offered a high-quality environment, then they would feel respected and be respectful. Matters such as non-matching taps would detract from this impression of a high-quality environment.

At a similar level of detail, one school facilities manager complained that the contractors had used narrow ledged supports for ceiling tiles rather the slightly more expensive wider tracks. The use of narrow tracks resulted in ceiling tiles that were very easy to displace and this had become something of a focus for those students who wished to engage in some form of rebellious behaviour. There were many other examples of reduction in specification of materials resulting in lack of durability and erosion of the aesthetics of the school.

This reduction in the technical specification of the construction was very overt in both heating systems and ventilation control systems. In the case of all the biomass boiler systems in schools we visited, this made them unfit for purpose and necessitated costly replacement.

School design and standardisation

One of the aims of the *Design Matters?* project was to make recommendations regarding the desirable level of standardisation in school design. This aim was formulated in the context of the UK government's decision to replace BSF with a much lower cost, more tightly regulated school building programme. As outlined in Chapter 2, the Priority School Building Programme (PSBP) was established in 2011 to rebuild, or meet the condition needs of those schools that remained in the worst condition.

Taken together with the research of others (Barrett, 2015; Mumovic et al., 2009; Williams et al., 2015), we are confident that there are certain environmental features of design that are generally conducive to a healthy working environment, including clean air, consistent warmth, level lighting and good acoustics. At the whole-school level, good sightlines and lack of tight corridors, stairways and 'bottlenecks' is helpful in reducing bullying and making students – particularly younger students – feel more secure. We note Barnett's argument that certain colour schemes are more positive than others but consider that, at this level of detail, there is insufficient evidence to impose standards. Clearly, there is scope for much more research in relation to these particular matters.

We are, however, unhappy about over-standardisation of school design. This is because we do not wish to rule out innovation, as this has impacts not merely on design but also on pedagogy. That research shows one existing design as outperforming others is not enough to render that design the 'last word'. Also, it is clear that different occupiers of schools can have somewhat different priorities and this, coupled with changes in schools policy, renders any ossification of the system as potentially counter-productive. Approval of new school designs should certainly be monitored to ensure that high environmental standards, including acoustic standards, are being maintained, however. Also, monitoring should ensure that significant areas are not being created that are likely to be little utilised.

Despite our unwillingness to endorse narrow standardisation of specific design features, we welcome moves to bring greater order to the process of commissioning and building new schools, whatever their design. As detailed in Chapter 4, we have found evidence of great inefficiency in certain cases, in terms of both human and financial resource.

We now move to a brief overview of the final element of our empirical work which aimed to find out how the generation of new builds we studied are faring over a longer period.

The national picture: an update

Our research has led us to seek a broader picture on some of our AHRC findings about the implications of occupying school buildings over an extended period of time and through successive changes in educational practices and leadership for a larger sample of schools. The *Design Matters?* project was an example of in-depth pedagogic post-occupancy evaluation. At the very end of the project and just before submitting this text for publication we conducted a national telephone and email survey of all schools built under the BSF programme. The findings of this small national BSF study are summarised below.

Methodology

BSF was a £55 billion, 15-year programme to rebuild or renovate all secondary schools in England that was cancelled after six years. On 19 July 2010, the Department for Education published an updated list of 1,500 schools included in the BSF programme stating whether the building project was now 'open', 'unaffected', 'under discussion' or 'stopped'. The list was produced by Partnerships for Schools and released by the DfE in July 2010 (DfE, 2010). Schools were classified in waves from pathfinder 1–15, local authorities were approved to commission new school buildings in successive waves from 2005 to 2010 (DfE, 2010). This classification helped us to explore potential differences between schools designed and built at different times. Some larger local authorities were in more than one wave. We classified schools by the first date that the local authority joined the programme. Projects that have been labelled as 'stopped' and ICT only, and those that have been confirmed as closed based on our desk research were excluded from this study.

A total of 602 schools were contacted by email and by telephone between November 2017 and January 2018. Three questions were asked:

1. Have you made any modifications to the building since moving in? If yes, what kind of modifications?
2. Did your design have enclosed classrooms, open-plan learning spaces, or both? Any comments?
3. Now you have been in the building for some time, is there anything you would like to change about it? If yes, what changes would you make?

Additional follow-up questions were asked, as appropriate, with each participating school during telephone interviews with facilities managers or members of the senior leader team. The result was 239 schools participating in either the email survey or the telephone interview, which is a small sample but indicates an emerging pattern for BSF schools in occupation through an extended period of time.

Data analysis

We analysed the types of modifications that had been untaken and the reasons that were given for undertaking these modifications. Sixty-two per cent of the participating schools reported that they had made modifications to the building since occupation; 74 per cent of the participating schools reported that they would like to make changes in the future.

Design in practice

Sixty-three per cent of the schools that we interviewed have an element of open-plan learning spaces (including schools with open-plan learning space only or a mixture of open-plan and enclosed learning spaces). Schools with open-plan learning spaces were more likely to have made modifications to their school buildings (71 per cent compared to 42 per cent), or would like to make modifications in the future (74 per cent, compared to 50 per cent). The data shows that the 83 per cent of the schools that had made modifications to their buildings since occupation were schools in the earlier waves – pathfinder 1, 2 or 3. The major modifications reported were all to their open-plan learning spaces by changing them into enclosed classrooms. Seventy-six per cent of the comments from schools with open-plan learning spaces mentioned dissatisfaction with the open-plan learning spaces:

> We simply can't use the space for learning and teaching.
> The open zones are too difficult to teach in.
>
> *(Headteacher)*

178 What matters about design?

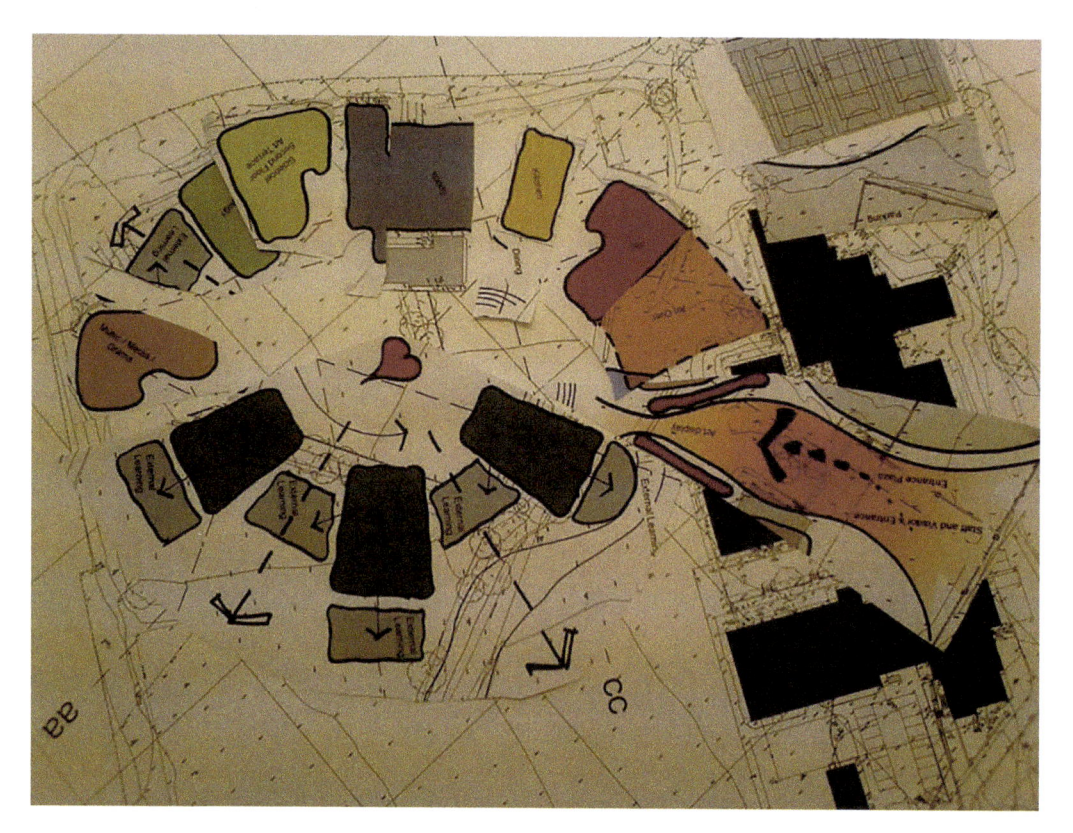

FIGURE 8.1 Concept diagram of case study school
Credit: HKS Architects

The schools reported that their open-plan learning spaces were either enclosed or the spaces were used for purposes other than large group teaching and learning as envisaged in the original design. The relationship between the changing pedagogical practice and design were often reported:

> A lot of our open-plan classrooms have been enclosed as teachers found it very difficult to teach in. However, they do make the school feel open and airy which is what makes our school different and appealing. The use of open classrooms should be considered very carefully with an eye to the future as requirements in school change very, very quickly.
>
> *(Headteacher)*

The schools that reported satisfaction with their designs described how their school community were 'heavily involved in the design of the school' and how the design matched their pedagogic needs. Schools used their open-plan learning areas for a range of activities from large group teaching to breakout, student group study areas and for group intervention. The data shows that a lower percentage of schools in subsequent Waves 4–15 (33 per cent) have an element of open-plan learning spaces. The majority of Waves 4–15 schools had enclosed classrooms only for teaching. In addition, 28 per cent of schools surveyed in Waves 4–15 reported that the enclosed classrooms were too small for the class size and did not allow for flexible configurations for group work or a variety of seating plans.

Flexibility and adaptability

In BSF guidance it was suggested that 'flexible or agile designs will, first, accommodate the day-to-day changes in pedagogy, and second, accommodate for long-term expansion or contraction' (Building Futures, 2004). The need for flexible designs to accommodate the day-to-day changes in pedagogy and longer-term expansion were reported by many schools we interviewed.

Twenty-eight per cent of schools reported that they have extended the school buildings, 34 per cent of schools reported that they would like to make future expansions. These related to different areas of the school, larger classrooms as well as playgrounds, storage, staircases, dining halls, PE halls, staff meeting rooms and toilets. Many schools reported congestion issues that related to movement and circulation around schools which impacted on behaviour, staff monitoring, noise and timetabling for lessons, lunch and breaktimes.

Feelings of safety

According to BSF guidance, school security, was concerned with 'creating a feeling of a secure, organised, safe environment' (Partnerships for Schools and 4ps, 2008; OECD, 2006; CABE, 2007).

Sixteen per cent of schools reported safety-related concerns. Many schools reported on the effectiveness of open, high visibility designs on passive supervision and behaviour:

> It's easy to pick up a problem quickly, the design is so open, we can see every student. There are no hiding places in this school. The students feel really safe.
>
> *(Headteacher)*

In addition, schools discussed how visual connectivity support a 'sense of being part of the school community as you walk around the school' (Headteacher).

Schools also reported on the disadvantages of areas without passive supervision:

> There is an area of the school where performing arts and PE spaces are located where we have traditional corridors with no passive supervision and this is where we find behaviour can be less good.
>
> *(Headteacher)*

Build quality, comfort and sustainability

Build Quality, according to DQI (2014), concerns how well the building is constructed: its structure, fabric, finishes and fittings, its engineering systems, the co-ordination of all these and how well they perform. Comfort and sustainability are closely connected to build quality.

Thirty-three per cent of schools reported concerns with build quality, maintenance and sustainability. Many schools are concerned with issues related to ventilation, natural daylight, thermal comfort, acoustic, and ongoing maintenance problems such as leaking roof, broken boiler, broken water tank, and bad quality door handles that they struggle to resolve. Build quality issues have a direct impact on the everyday experiences of students and teachers and are shown to have detrimental effects on concentration, mood, well-being, attendance and attainment (e.g. Bluyssen, 2017; Gunter and Shao, 2016; Barrett et al., 2015):

> We have no problems with the overall layout of the building, but there are problems with quality, the floor in the theatre is bubbling up and the roof is leaking in many places. The builders were contacted and asked to come back, but they never did.
>
> *(Headteacher)*

Typical post-occupancy evaluations tends to narrowly focus on environmental factors (e.g. Winterbottom and Wilkins, 2009), yet 48 per cent of schools reported concerns about the environmental performance of their buildings. Overwhelmingly, schools reported limited or lack of local control for students and teachers, overcomplicated building management systems that require external assistance, maintenance, poor performance in sustainability in practice and increased energy consumption.

The impact of PFI

The Department of Education announced in 2004 that of the £2.2 billion funding for the first wave of BSF, £1.2 billion (55 per cent) will be covered by Public Finance Initiative credits (PFI) (http://web.archive.org/web/20080924172144/www.number10.gov.uk/Page5801) where the buildings were effectively constructed and maintained by a private company, then leased back to the local government for a guaranteed period. Eighty-five per cent of PFI schools in our survey reported that they are struggling with the financial and operational impacts of the PFI contracts:

> PFI costs are going up every year. Modifications are very restricted, it takes a long time to negotiate with facilities and site managers. Modifications also cost a lot more within the contract, there is a 20 per cent extra on everything we have to buy, also an extra 8 per cent in every negotiation phase.
> *(Headteacher)*

Schools reported that control has been taken away from the schools through the PFI contracts. Schools are tied to one construction company for maintenance and facilities management. Eighty-one per cent of PFI schools said they would like to make modifications to the building, however, they are restricted in many ways due to the nature of the contract:

> The school building is not ours. We have no control over anything and maintaining the building is very expensive. To buy something that is £50, we actually have to spend £80–100 due to the PFI contract.
> *(Headteacher)*

There are also major concerns about the quality of maintenance care and modifications:

> The school was built for the whole community but with these prices no one in the community can afford to use the building, outside of school hours we cannot afford to have extra clubs for the students. Before if we had any issues with the building, somebody on site would just fix it. Now we have to do everything through the phone and nothing is ever fixed properly.
> *(Headteacher)*

While much of previous literature discussed the criticism of PFI provision in general within the public sector (Pollock et al., 2002), there is a lack of more specific studies drawn from the experiences of school buildings in occupation through time.

SEN and inclusion

Special Schools for students with special educational needs and pupil referral units reported higher levels of concern on the fitness for purpose of their school designs. Eighty-five per cent of PFI schools said they

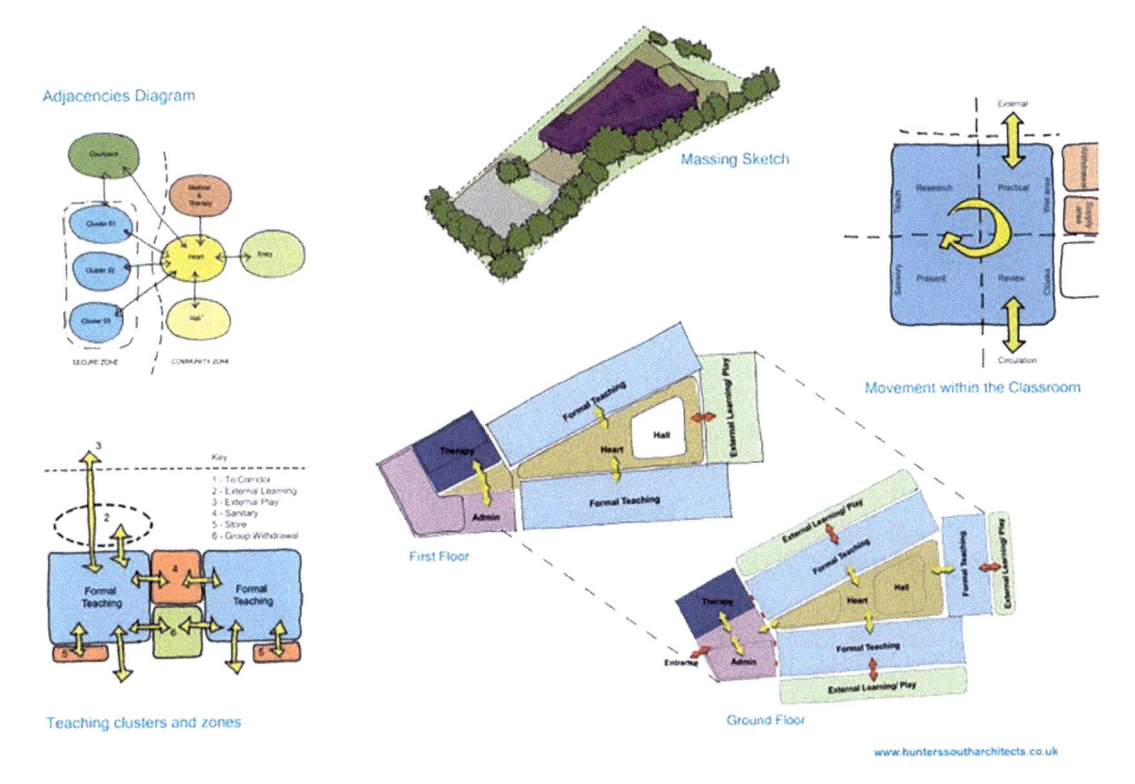

FIGURE 8.2 Concept drawing of special school
Credit: Hunters South Architects

would like to make modifications to the design but many reported the challenges of finding funding to make their buildings fit for purpose. Many schools also reported concerns about the unmet needs specific to their students with SEN and issues around safeguarding:

> This is a school for students with special needs, but it is not reflected in the design of the building, the doors and the furniture are too weak, windows are not safe, students can get out easily, the fire extinguishers are not hidden, they are easily available for students. The school was built with the specifications of a mainstream school without considering the special needs of our students.

Research on inclusive school design most typically focuses on a narrow range of children, children with physical difficulties (e.g. Mihaylov et al., 2004; Law et al., 2007), and those with autistic spectrum disorder (e.g. Martin, 2016; McAllister and Sloan, 2016; Scott, 2009). Other groups such as those with moderate learning difficulties, challenging behaviour, or those deemed to be vulnerable or at risk are poorly represented, despite forming a larger proportion of the school population. Focusing exclusively on specific groups does not facilitate an understanding of their competing needs in a school environment (Clark, 2002).

Discussion

> It's very difficult, there are many things we can't afford to fix. The design looks good but doesn't function for us.
> *(Headteacher)*

The findings from this small sample of BSF schools in occupation correlates with the findings from *Design Matters?*, the central issue is obviously not as simple as whether classrooms are designed to be open or enclosed but whether school designs can support a range of different pedagogic approaches through time. If school designs can be 'tools for teaching and learning', then design practitioners require a deeper understanding of the mutual shaping of design and pedagogic development.

The current UK government has recognised the urgent need for more new schools at a time when major concerns have been expressed about the school estate (House of Commons Committee of Public Accounts, 2017). The Free School Programme (FSP) (currently more than 600 schools) was instigated alongside the £4.4 billion Priority School Building Programme (PSBP) for the rebuilding and refurbishing of school buildings in the worst condition. The Education and Skills Funding Agency's (ESFA) funding for school buildings is now at £1,113/m² reduced by approximately a third from those incurred during BSF (ESFA, 2014; www.gov.uk/government/publications/baseline-designs-for-schools-guidance/baseline-designs-for-schools-guidance). Project time has also been reduced from 24–36 months to 12 months to drive efficiency. This involves significantly limiting consultation with school communities and multiple stakeholders. The Education Funding Agency (EFA) has produced Control Options in order to demonstrate how their standardised baseline designs should be applied in practice. There is still very limited research on the pedagogic implications of these standardised school designs and their fitness for the purposes of occupiers. Below is a summary of the key lessons learned from *Design Matters?* that can begin to help us understand the complexities of the interplay between school design and pedagogic practice.

Summary of findings

This final section will summarise what we feel we have learned from the *Design Matters?* project overall, taking together the various forms of data analysis, qualitative and quantitative, combined with the 'noticings' discussed earlier in this chapter. These are presented under three headings: design and practice, processes of learning, and pedagogic post-occupancy evaluations.

Design and practice

Design and practice are mutually interactive. While it is tempting, as OECD do, to talk about design and practice 'shaping' each other (OECD, 2014), this can also be construed as an over-strong claim, as 'shaping' suggests some causation. Our research suggests that the primary driver here is practice and expectations related to practice. This explains why some headteachers have worked positively with the BSF designs while others have effectively undermined them. Sometimes buildings are seen as fit for purpose; if they are not, they do not of themselves tend to change one's mind about purpose.

For example, when new leaders enter their adopted school buildings for the first time, it is not as a blank slate, to use Locke's term (Locke, 1959). They will know what they want to achieve with the school; indeed, they will have been appointed on that basis. In other words, the new senior management team will enter the building with a certain form of practice in mind. The building will then seem to offer certain opportunities, or invitations, and certain restraints, which will then modify the resulting practice to some extent, but not determine it.

We found strong evidence that one design may be perceived and used in very different ways in different practices of schooling. Furthermore, over time a process of accommodation happens whereby there is some mutual effect of design and practice on each other.

It is certainly the case that lack of congruence between design and practice can result in space which is not fit for purpose. Different approaches to school leadership and management give rise to distinctive school cultures which in turn make differences in the use and adaption of a school building. One of the outcomes of the project was the realisation of just how often any school changes not only leaders but leadership style. This is often in response to poor inspection reports and unsatisfactory results, and is most likely to happen most frequently in areas of long-term low achievement, such as those in which many of the BSF flagship schools were built. Of our original five main project schools, only one has kept the same headteacher since the beginning of the project while three have changed leadership on more than one occasion during a five-year period.

Processes of learning

There is an increase in research interest in the relationship between design and educational outcomes. Perhaps the strongest evidence so far comes from Barrett's recent studies of classrooms, albeit there may be some ambiguity in Barrett's research regarding design and teacher effects. Barrett claims that design can contribute 16 per cent of differences in outcomes (Barrett et al., 2015)

Barrett's work relates to classrooms, not schools as a whole. It is difficult to find such strong evidence for design at the whole-school level increasing outcomes other than over the short term; indeed, Williams et al. (2015) have found that schools tend to achieve improved results even before the new building is completed, so there is some kind of halo effect at play here. This may be particularly the case with schools in challenging circumstances, such as those forming the main sample in the *Design Matters?* project. Indeed, we found no consistent pattern of increased performance in direct relation to school design. On the other hand, these were acknowledged as challenging schools and they also changed hands frequently as management teams were replaced. In line with Williams et al. (2015), we found a consistent but short-term boost in student – school connectedness after occupation in the BSF designs, but this was neither maintained over the long term nor obviously connected to increased educational outcomes.

There is an argument to the effect that new occupiers of schools should respond more sensitively to the opportunities that the new building has to offer. It is certainly the case that design can provide opportunities for pedagogic change but a process of occupation is required to prepare users for a new kind of *work*.

Pedagogic post-occupancy evaluations

The way school buildings and estates are currently evaluated is inadequate. Post-occupancy evaluations, as traditionally practised, are too narrowly focused on easy measurable aspects of building performance such as energy use. These matters are important; indeed, a well-lit, well-ventilated classroom can help learners and teachers (Barrett et al., 2015), not least by reducing levels of air pollution (Mumovic et al., 2009). However, they do not take enough account of users, whose motivations and expectations related to the building may differ.

The most difficult challenge for designers comes in aiding rather than abetting the process by which a space can become first a place, then a 'home'. Of course, 'homes' are created ultimately by occupiers, through their practices and relationships: an architect cannot ensure a sense of home. However – and particularly, perhaps, in the case of schools – our research reveals that design can help to enhance or hinder social relations, and social relations are of central importance to schools. A school design that can enhance social relations is, in the most important single respect, a good school design, since social relations cover areas as basic as teaching and learning, personal security and well-being, and peer relations. Design may help to reduce bullying, promote a sense of calm and personal space, pride in the school or a sense of alienation. Post-occupancy evaluations should be designed in order to uncover how this works in differing contexts and with differing student and teacher populations.

There is, nevertheless, an issue relating to variability of context and the useful life of a school: an issue of what constitutes sustainability in school design. Even over the five-year period between the beginning of the *Design Matters?* project and the time of writing, schools have changed not only leadership but also management and pedagogic style on several occasions. How can one design, therefore, be fit for 'the future'? There is no easy answer to this dilemma. Our in-depth research on a small number of BSF and Academy schools suggests that not all original design intentions will be realised in practice for very long in most cases, particularly when fundamental changes take place in educational policy with changes in government.

Overall, our findings suggest a need to redefine 'sustainability' in terms of adaptation to different forms of practice. In order to extend the functional life of new school buildings, the vision and design process must allow for adaptation as educational policies and practices change through time. In practice, this means that school design requires good multi-professional holistic post-occupancy evaluation which has a remit that goes far beyond the physical/environmental functioning of the building.

Above all, an understanding of the social relations that are enacted within a design as it is taken up by different forms of practice is crucial to the development of better sites for schooling.

References

Barrett, P., Zhang, Y., Davies, F. and Barrett, L. (2015) *Clever Classrooms: Summary Report of the HEAD Project*. Salford: University of Salford.

Bluyssen, P.M. (2017). Health, comfort and performance of children in classrooms – new directions for research, *Indoor and Built Environment*, 26(8): 1040–50.

Building Futures (2004) 21st century schools: Learning environments of the future. http://webarchive.nationalarchives.gov.uk/20110118205716/www.cabe.org.uk/files/21tt-century-schools.pdf

CABE (2007) Creating excellent secondary schools: A guide for clients. http://webarchive.nationalarchives.gov.uk/20110118111850/www.cabe.org.uk/files/creating-excellent-secondary-schools.pdf

Clark, H. (2002) *Building Education: The Role of the Physical Environment in Enhancing Teaching and Research*. London: Institute of Education.

Cooper, I. (1985) Teachers' assessments of primary school buildings: The role of the physical environment in education, *British Educational Research Journal*, 11(3): 253–69.

Daniels, H., Tse, H.M., Stables, A. and Cox, S. (2018) Design as a social practice: The experience of new build schools. *Cambridge Journal of Education*, 43(6): 767–87.

Department for Education (2010) List of schools. http://news.bbc.co.uk/1/shared/bsp/hi/pdfs/19_07_10_school_error_list.pdf

Education and Skills Funding Agency (2014) Baseline designs for schools: Guidance. www.gov.uk/government/publications/baseline-designs-for-schools-guidance/baseline-designs-for-schools-guidance#cost-and-area-allowances

DQI (2014) www.dqi.org.uk/perch/resources/dqi-schools-guidance-copy.pdf

Gunter, T. and Shao, J. (2016). Synthesising the effect of building condition quality on academic performance, *Education Finance and Policy*, 11(1): 97–123.

House of Commons Committee of Public Accounts (2017) *Capital Funding for Schools*. London: HMSO.

Lave, J. and Wenger, E. (1991) *Situated Learning. Legitimate Peripheral Participation*. Cambridge: Cambridge University Press.

Law, M., Petrenchik, T., King, G. and Hurley, P. (2007) Perceived environmental barriers to recreational, community, and school participation for children and youth with physical disabilities. *Physical Medicine and Rehabilitation*, 88(12): 1636–42.

Locke, J. (1959) *An Essay Concerning Human Understanding*. New York: Dover Publications.

Martin, C.S. (2016) Exploring the impact of the design of the physical classroom environment on young children with autism spectrum disorder (ASD), *Journal of Research in Special Educational Needs*, 16: 280–98.

McAllister, K. and Sloan, S. (2016) Designed by the pupils for the pupils: an autism-friendly school, *British Journal of Special Education*, 43(4): 330–57.

Mihaylov, S.I., Jarvis, S.N., Colver, A.F. and Beresford, B. (2004) Identification and description of environmental factors that influence participation of children with cerebral palsy, *Developmental Medicine and Child Neurology*, 46(5): 299–304.

Mumovic, D., Palmer, J., Davies, M., Orme, M., Ridley, I., Oreszczyn, T., Judd, C., Critchlow, R., Medina, H.A., Pilmoor, G., Pearson, C. and Way, P. (2009). Winter indoor air quality, thermal comfort and acoustic performance of newly built secondary schools in England, *Build. Environ.*, 44(7): 1466–77.

OECD (2006) *The Case for 21st Century Learning*. Paris: OECD. www.oecd.org/general/thecasefor21st-centurylearning.htm

OECD (2014) *Effectiveness, Efficiency and Sufficiency: An OECD Framework for a Physical Learning Environments Module*. Paris: OECD.

Partnership for Schools and 4ps (2008) *An Introduction to Building Schools for the Future*. London: Department for Children, Schools and Families.

Pollock A.M., Shaou, J. and Vickers, N. (2002). Private finance and 'value for money' in NHS hospitals: A policy in search of a rationale? *BMJ*, 324(7347): 1205–9.

Price, I., Clark, E., Holland, M., Emerton, C. and Wolstenholme, C. (2009) *Condition Matters: Pupil Voices on the Design and Condition of Secondary Schools*. Reading: CfBT Education Trust.

Scott, I. (2009) Designing learning spaces for children on the autism spectrum, *Good Autism Practice*, 10(1): 36–51.

Sharp, R. and Green, G. (1975), *Education and Social Control: A Study in Progressive Primary Education*. London: Routledge and Kegan Paul.

Tse, H.M., Learoyd-Smith, S., Stables, A. and Daniels, H. (2014). Continuity and conflict in school design: A case study from Building Schools for the Future, *Intelligent Buildings International*, 7(2–3): 64–82.

Williams, J.J., Hong, S.M., Mumovic, D. and Taylor, I. (2015) Using a unified school database to understand the effect of new school building on school performance in England, *Intelligent Buildings International*, 7(2–3): 83–100.

Winterbottom, M. and Wilkins, A. (2009) Lighting and discomfort in the classroom, *Journal of Environmental Psychology*, 29(1): 63–75.

APPENDIX

School connectedness questionnaire (adapted)

How do you feel about school?

These questions are about how you feel about school. Tick a box to show how true each statement is for you.

	Not at all true	Hardly ever true	Sometimes True	Often True	Completely True
1. I feel like a real part of this school	☐	☐	☐	☐	☐
2. People at this school are friendly to me	☐	☐	☐	☐	☐
3. I am treated with as much respect as other students	☐	☐	☐	☐	☐
4. I can really be myself at this school	☐	☐	☐	☐	☐
5. The teachers here respect me	☐	☐	☐	☐	☐
6. I feel proud of belonging to this school	☐	☐	☐	☐	☐
7. I feel safe in this school during lessons	☐	☐	☐	☐	☐
8. I feel safe in this school during break/lunchtime	☐	☐	☐	☐	☐
9. It is easy to find my way around this school	☐	☐	☐	☐	☐
10. There are lots of places for me to be with my friends in this school	☐	☐	☐	☐	☐
11. I can have time on my own in this school if I want to	☐	☐	☐	☐	☐

INDEX

Page numbers in *italics* refer to figures and tables.

Academies Programme 7, 10, 90, 140
acoustic consultant 78–80, 81
acoustics: and open-plan spaces 109, 110, 134; value engineering and cost saving 175
activity theory 33–4, 62
adaptability and flexibility 169–71, 179
architects 74–6, 77, 102–4, 131; and academic educationalist collaboration 33; and contractor roles 114, 171; delivery 76–7, 78, 80, 97, 109; interviews 76–7, 80, 90–2, 93, 94, 102–4, 109, 131
Arts and Humanities Research Council (AHRC) 34, 176

Baars, S. et al. 35, 61
Barrett, P. et al. 16, 34, 183
Bazerman, C. 64
behaviour/behaviour change 161; aggression 101, 172; control 148; exclusion 129; vandalism 94, 175–6
belonging 123–9; engagement and appropriation 172–4
Bernstein, B. 139, 140, 141–4, 145–6, 162, 163
Bexley Academy 90–4
biosemiotics 42
Blair, T. 90
boundaries 174
Bourdieu, P. 3, 42
breakout spaces 74–5, 92, 99, 109, *110*, 113, 131, 168, 169
British Council for School Environments (BCSE) 25
build quality 175–6, 179–80
Building Bulletin (BB) 23, 74, 78–9, 81–2, 140

building information modelling (BIM) 23
Building Schools for the Future (BSF) 7, 8–14, 18, 22, 35, 68–70, 94, 95–7, 122, 140; *see also* Department for Education and Skills (DfES)
Burke, C. 44; and Grosvenor, I. 25

CABE *see* Commission for Architecture and the Built Environment
Cartwright, N. 73; and Hardie, J. 16
Carvalho, L. and Dong, A. 7, 140
changing schools (primary to secondary) 139–40; comparator school analysis 150; design/practice alignment analysis 151–2; group experiences of transition 152–5; individual experiences of transition 155–61; methodology 144–9; modalities and analysis 146–9, 150–1, 155–61; policy context 140; research findings 149–52; school-level data and analysis 145–9, 149–50; student-level data and analysis (school connectedness questionnaire), 144–5, 149–52, *154*, *158*, *160*, 187 *appendix*; summary and conclusion 161–3; theoretical framework 140–4; *see also* new-build schools
child development 121
child protection 51; and open community school 169
classification 142, 144, 146–9; and coding 145–6
Cleveland, B. and Fisher, K. 34–5, 61; and Soccio, P. 61
co-configuration 13, 64, 65, 97, 98, 112, 115
Cole, M. 1–2, 121, 136; Griffin, P. and 121
collaboration 1, 8–9, 12–14, 33, 114–15; building information modelling (BIM) 23; and personalisation 95
Commission for Architecture and the Built Environment (CABE) 8, 22, 68, 69, 72, 73, 76, 80

Index

communication: competitive bidding process 76–7; coordination, cooperation and 62, *63*; end-users and professionals 21–2; problems 104, 109; stakeholders 12; ZPD 64
community cohesion 93
community engagement 90–4
community and families 148–9
competence 121, 130, 131; instructional and regulatory discourses 142–3
competitive bidding/dialogue process 76–7, 97, 103
conflict: and continuity 67–83, 134–5; frustration and dissatisfaction 97–8, 102–6, 172
construction industry: critiques 12, 23; limited research on 14, 23; tools and collaboration 23
construction materials 175–6, 179–80
consultation 9, 11, 21–2; process 51, 72–3, 75, 76, 77, 81, 83, 93–4, 97, 113
contextualising and recontextualising design and building 68–74
Cooper, I. 21, 25, 61, 172; Foxell, S. and 16–17, 34
coordination, cooperation and communication 62, *63*
costs: consultants' fees 76; Control Options (EFA/ESFA) 11, 182; design and build stage 78, 80, 133; repainting 94; sponsors 90; value engineering 97–8, 175
councils: policy to individual design 98–113; *see also* Local Authority Transformation Team
cultural transmission: semiotic account 142–3, 162–3; wall display study 36

Daniels, H. 1, 33, 36, 122, 139, 142; et al. 84, 120, 121, 123, 136, 163, 170
delivery architects 76–7, 78, 80, 97, 109
demographic changes 168
Den Besten, O. et al. 22, 76
Department for Children, Schools and Families (DCSF) 8
Department for Education (DfE) 129, 140, 177, 180
Department for Education and Skills (DfES) 9–10, 18, 19, 68, 71–2, 73, 74, 95; *see also* Building Schools for the Future (BSF)
design and build stage 77–8
design and building team 76–7
Design Matters? project 33–4; aims 45; data analysis 55; instruments 53–5; methodological considerations and implications 44–5; methodology 46, *47*; origins 36–40; overall schema *2*; overview 1–3; phases 51–2; philosophical and research problem 34–5; research questions 45–6, 51–2; sample 46–51; semiotics 41–4; space to place challenge 35; summary of findings 182–4
design and practice: alignment analysis 151–2; Locality A 90–4; Locality C 94, 95–8, 98–113; relationship between 7, 14–24; summary and conclusions 113–17; summary of findings 182–3
disabilities *see* special educational needs (SEN)
discourses: instructional and regulatory 142–3, 144, 146–7; overlapping 43

display work *see* wall display work
documentary analysis 55

Earthman, G.I. 14
Education Funding Agency/Education and Skills Funding Agency (EFA/ESFA) 11, 182
educational vision 12, 72–4; transforming into design 74–6
edusemiotics 2, 42
Egan Report 12
engagement: belonging and appropriation 172–4; community 90–4; user 26
Engineering and Physical Science Research Council (EPSRC) 34
English language support 168
Engström, Y. 13; and Ahonen, H. 13; et al. 62, 64; and Middleton, D. 13
essay task 44–5, 54
ethnic mix in schools 168
evidence-based policies 73
evidence-based practice 16–17
exclusion 129
External Relations as Enacted in Design 145, 147

field and habitus 3, 42
finance/funding *see* costs
flexibility and adaptability 169–71, 179
formal classroom observations 54
Foxell, S. and Cooper, I. 16–17, 34
Free School Programme (FSP) 182

Gearhart, M. and Newman, D. 36–7
Gieryn, T.F. 17, 23
Goodenow, C. 53–4, 66, 145
Gove, M. 11, 70
governance 163
Government Soft Landings (GSL) 23
Greene, J.C. 55, 66
Griffin, P. and Cole, M. 121

habitus and field 3, 42
headteachers (including interviews) 171, 172, 173, 177–8, 179, 180, 182; Modality A and B 147–9; new-build schools 161; School C1 98–102, 129–30; School C2 102–3, 104–6, 131–2; School C3 106–9, 133; School C4 110; School C5 112–13; Secondary Transformation Team (STT) 74
'heart space' designs 46, 51, 107, 112, 123, 126–8, 129
hierarchy/power relations 142–3, 144
Higgins, S. et al. 14–16, 21
Hillier, H. and Hanson, J. 18
Hoadley, U. 142, 143–4
House of Commons: Education and Skills Committee 68, 69, 70
House of Lords Select Committee on National Policy for the Built Environment 13–14
Hundeide, K. 162

ICT 18–19, 20, 68, 96; technological 'future' 167–8
inclusive practice 19–20, 180–1
informal observations 54
Innovative Learning Environments (ILE) 60
inquiry-based learning 96
instructional and regulatory discourses 142–3, 144, 146–7
interviews 54, 66, 67; architects 76–7, 80, 90–2, 93, 94, 102–4, 109, 131; Secondary Transformation Team (STT) 69, 72, 73, 74, 76, 81, 96; students 130, 132, 134, 135; teachers 130, 132, 133; *see also* headteachers
Ivic, I. 141

James Report/Review 11, 22–3, 69–70, 72, 95, 97

The Key study 12
knotworking 13, 64, 65, 97
knowledge, types of 64–5
Knowledge Transfer Champions 34
Kraftl, P. 10, 11, 68, 122

Labour Governments 68, 72, 90, 95
Latham Report 12
leadership 9–10, 134, 182–3; *see also* headteachers
learning environments 35, 59, 60
learning processes 183
learning 'villages' 90–2, *93*
literacy, types of 43
Livesey, P. 70
local authority policy to individual design 98–113
Local Authority Transformation Team 102, 122, 131, 133, 136, 170
local education partnerships (LEPs) 8
local recontextualisation of national policy 95–8

Mahoney, P. et al. 68, 69; and Hextal, I. 68
Mandelson, P. 68
map task 53
mediation, concept of 120–1, 130–1, 141–2
'messages' 43–4
Microsoft Consulting Services 18
Milliband, D. 18, 71, 95
mixed economy of space 9, *15*, 20–1, 99, 131–2, 144, 147–8
Morrell, P. 12, 23
'Mosaic' approach 26
multi-agency/multi-professional approach 62, 64, 69, 81, 84, 170
'My First Week at This School' (essay task) 44–5, 54

National Audit Office (NAO) 69, 74, 76, 83, 95, 97, 119–20, 136
national BSF programme study 176–7; data analysis and findings 177–82; methodology 177
national policy 140; and joined-up thinking 169; local recontextualisation 95–8; *see also* Building Schools for the Future (BSF); Priority School Building Programme (PSBP)

networking *see* knotworking
'new cathedrals of learning' 9, 140
new-build schools 119–20; case studies 47–51, 129–35; pedagogic post-occupancy evaluation and methodology 122–3; social space, importance of 128–9; student perceptions 123–8; summary and conclusion 135–6; theoretical orientation 120–2
Newman, D. et al. 64; Gearhart, M. and 36–7
nominal group techniques (NGTs) 44, 52, 53, 66, 122–3, 133
nursery school children 36–7

observational methods 54–5
occupation stage 80
OECD 16, 17, 21, 59, 60, 95, 182
Ofsted report 93
open learning zones 130, 131–2, 135; acoustics 109, 110, 134; retrofit 133–4; 'schools within a school' model 106–7, 133
open-plan learning spaces 20–1, 79–80, 81–2, 96, 97, 101, 102–3, 104–6, 112, 177–8; and traditional classroom design 14–16, 21, 105–6, 131–2
openness and control 173–4

parents 9, 52, 90, 173–4; community and families 148–9
Parnell, R. et al. 12
participatory design process 60
Partnerships for Schools (P*f*S) 8, 69, 177, 179
Pedagogic Practice as Enacted in Design 145, 146–7
pedagogical practices 18
personalisation/personalised learning 9, 13, 95, 140
philosophical and research problem 34–5
photographs and informal observations 54
Piaget, J. 143
post-occupancy evaluation (POE) 25–6, 34–5, 41, 59–60, 114–15, 167; broadening 171–2; multi-professional 170; pedagogic 61–2, 122–3, 174, 183–4; stage model 80
power relations/hierarchy 142–3, 144
prayer room 168
primary schools 16; display work 37, *37–40*; transition to secondary schools *see* changing schools; new-build schools
Priority School Building Programme (PSBP) 11, 22, 23–4, 70, 182
Private Finance Initiative (PFI) 68, 69, 70, 180
professional action/inter-professional working 13–14
project architects 102–4, 131
'prolepsis'/proleptic instruction 121, 132, 136
psychosocial learning environment (PSLE) and physical learning environment (PLE) 35

recontextualisation: design and building 70–4; policy 95–8
regulatory and instructional discourses 142–3, 144, 146–7
Reid, K.C. and Stone, A. 121–2

religion: prayer room 168
renovation/refurbishment 22, 131
retrofit 133–4
RIBA 10, 80
Rutter, M. 14

safety 179
Sailer, K. and Penn, A. 60–1
school connectedness questionnaire 53–4, 144–5, 149–52, *154*, *158*, *160*, 187*a*
school population and classroom size 168
School Spaces Evaluations Instrument (SSEI) 61
school-building programmes 68–9
Schools for the Future 71
Schools Improvement Partner Programme 95–6
'schools within a school' model 106–7, 133
Secondary Transformation Team (STT) 72, 73, 74, 80, 95–6; interviews 69, 72, 73, 74, 76, 81, 96
self-assessment 13
semiotics 2–3, 33–4, 142–3, 162–3; codes and interpretations as readings 42–4; signifying environment (*umwelt*), school as 41–2; signs/signification (wall displays) 36, 41
Sigurðardóttir, A.K. and Hjartson, T. 14, 60
social class issues 171–2
social practice of design 59–60, 89; data analysis 67; methodology 62–6; pedagogic post-occupancy evaluation (POE) 61–2; research questions 66; stage model 67–83; summary and conclusion 83–4; theoretical orientation 60–1
social relations and buildings 17–18
social space, importance of 128–9
space to place 35, 80, 172
special educational needs (SEN) 71; buildings fit for purpose 23–4; inclusive practice 19–20, 180–1
Stables, A. 1, 2–3, 33, 42; et al. 43; and Semetsky, I. 42
stakeholders, communication between 12
standardisation 176; mass customisation 65–6

'Strategy for Change'(S*f*C) 69
students: behaviour *see* behaviour/behaviour change; first person research instruments 53–4; interviews 130, 132, 134, 135; nominal group techniques (NGTs) 44, 52, 53, 66, 122–3, 133; perceptions 123–8; school connectedness questionnaire 144–5, 149–52, *154*, *158*, *160*, 187*a*; self-perceptions 25; and teachers: third person approaches 54–5
sustainability 9, 115, 116–17, 169–70, 171, 179–80, 184

teachers: interviews 130, 132, 133; professional challenges 21–2; and students: third person approaches 54–5; *see also* headteachers
technological 'future' 167–8
Template for Schools of the Future 72–3, 96
traditional and open-plan designs 14–16, 21, 105–6, 131–2
Tse, H.M. 34; et al. 25, 120, 175
Tudge, J.R.H. and Winterhoff, P.A. 141

user engagement 26

value engineering processes 97–8, 175
Veloso, L. et al. 22
Victor, b. and Boynton, A. 64, 65–6, 115
vision: restructuring 10; *see also* educational vision
von Uexkhuell, J. 42, 45
Vygotsky, L.S. 2, 37, 41, 60–1, 62, 64, 120–1, 139, 140–2, 162, 163

Waddington, C.H. 115–17
wall display work 36–7, *37–40*; interpretation of 40–1
Wertsch, J.V. 37, 41, 142; et al. 1
Wittgenstein, L. 34, 42–3
Woolner, P. et al. 16, 21, 22, 25, 26, 59; and Thomas, U. 14

Zone of Proximal Development (ZPD) 64